D1781363

1/3/19

BUSINESS/SCIENCE/TECHNOLOGY DIVISION
CHICAGO PUBLIC LIBRARY
400 SOUTH STATE STREET
CHICAGO, IL 60605

Freedom in the Anthropocene

Other Palgrave Pivot titles

Christine J. Hong: **Identity, Youth, and Gender in the Korean American Christian Church**

Cenap Çakmak and Murat Ustaoğlu: **Post-Conflict Syrian State and Nation Building: Economic and Political Development**

Richard J. Arend: **Wicked Entrepreneurship: Defining the Basics of Entreponerology**

Rubén Arcos and Randolph H. Pherson (*editors*): **Intelligence Communication in the Digital Era: Transforming Security, Defence and Business**

Jane Chapman, Dan Ellin and Adam Sherif: **Comics, the Holocaust and Hiroshima**

AKM Ahsan Ullah, Mallik Akram Hossain and Kazi Maruful Islam: **Migration and Worker Fatalities Abroad**

Debra Reddin van Tuyll, Nancy McKenzie Dupont and Joseph R. Hayden: **Journalism in the Fallen Confederacy**

Michael Gardiner: **Time, Action and the Scottish Independence Referendum**

Tom Bristow: **The Anthropocene Lyric: An Affective Geography of Poetry, Person, Place**

Shepard Masocha: **Asylum Seekers, Social Work and Racism**

Michael Huxley: **The Dancer's World, 1920–1945: Modern Dancers and Their Practices Reconsidered**

Michael Longo and Philomena Murray: **Europe's Legitimacy Crisis: From Causes to Solutions**

Mark Lauchs, Andy Bain and Peter Bell: **Outlaw Motorcycle Gangs: A Theoretical Perspective**

Majid Yar: **Crime and the Imaginary of Disaster: Post-Apocalyptic Fictions and the Crisis of Social Order**

Sharon Hayes and Samantha Jeffries: **Romantic Terrorism: An Auto-Ethnography of Domestic Violence, Victimization and Survival**

Gideon Maas and Paul Jones: **Systemic Entrepreneurship: Contemporary Issues and Case Studies**

Surja Datta and Neil Oschlag-Michael: **Understanding and Managing IT Outsourcing: A Partnership Approach**

Keiichi Kubota and Hitoshi Takehara: **Reform and Price Discovery at the Tokyo Stock Exchange: From 1990 to 2012**

Emanuele Rossi and Rok Stepic: **Infrastructure Project Finance and Project Bonds in Europe**

Annalisa Furia: **The Foreign Aid Regime: Gift-Giving, States and Global Dis/Order**

palgrave▶pivot

Freedom in the Anthropocene: Twentieth-Century Helplessness in the Face of Climate Change

Alexander M. Stoner
Salisbury University, USA

and

Andony Melathopoulos
Dalhousie University, Canada

FREEDOM IN THE ANTHROPOCENE
Copyright © Alexander M. Stoner and Andony Melathopoulos, 2015.

All rights reserved.

First published in 2015 by
PALGRAVE MACMILLAN®
in the United States—a division of St. Martin's Press LLC,
175 Fifth Avenue, New York, NY 10010.

Where this book is distributed in the UK, Europe and the rest of the world, this is by Palgrave Macmillan, a division of Macmillan Publishers Limited, registered in England, company number 785998, of Houndmills, Basingstoke, Hampshire RG21 6XS.

Palgrave Macmillan is the global academic imprint of the above companies and has companies and representatives throughout the world.

Palgrave® and Macmillan® are registered trademarks in the United States, the United Kingdom, Europe and other countries.

ISBN: 978-1-137-50389-3 EPUB
ISBN: 978-1-137-50388-6 PDF
ISBN: 978-1-137-50387-9 Hardback

Library of Congress Cataloging-in-Publication Data is available from the Library of Congress.

A catalogue record of the book is available from the British Library.

First edition: 2015

www.palgrave.com/pivot

DOI: 10.1057/9781137503886

Contents

List of Illustrations	vii
Acknowledgments	viii
Prologue: Smog, Haze, Sulfur: The Elusive Clarity of the Anthropocene	1
1 The Great London Smog (1952)	2
2 Southeastern Asian Haze (2013)	6
3 Sulfur and the stratosphere (1815 and the future)	10
Introduction: What Is the Meaning of Freedom in the Anthropocene?	18
1 Georg Lukács (1885–1971) and the Critique of Reification: On the Dialectical Genesis of the Great Acceleration	28
1 Marxism in crisis: social democracy on the eve of the Great Acceleration	29
2 Critique of reification	33
3 Reification and the dialectical genesis of the Great Acceleration	42
2 Theodor W. Adorno (1903–1969) and the Critique of Identity Thinking: The Great Acceleration as Historical Sedimentation	48
1 Dialectics as critique	50
1.1 Critique of identity thinking	52
1.2 Negative philosophy of history	55

2 Human–ecological transformation and contemporary ecological subjectivity — 57
2.1 "Administered" society — 58
2.2 Toward a critique of contemporary environmentalism — 62

3 Moishe Postone (1942–) and the Critique of Traditional Marxism: Helplessness and the Present Moment of the Great Acceleration — 73
1 Critique of traditional Marxism — 74
2 Nature of the contradiction: value and material wealth — 79
3 Socioecological domination and the production of value — 82
4 Value as the continual necessity of the present — 86
5 Helplessness — 90

Conclusion — 99
1 Contemporary environmental politics — 103
2 The necessity of critical theory — 106

References — 110

Index — 121

List of Illustrations

Figure

0.1	Freedom and critique in the Anthropocene	24

Photos

A	The Great London Smog	2
B	Southeastern Asian Haze	7
C	Volcanoes and the stratosphere	11
0.1	The Anthropocene	19
1.1	Rosa Luxemburg and the Second International	31
2.1	Cold War technology and environmental subjectivity	63
2.2	Earth day and the environmental movement	67

Acknowledgments

The decision to write a book on the meaning of freedom in the Anthropocene was motivated by our felt need to expand upon and write about a panel series we helped develop for the Platypus Affiliated Society titled "Freedom in the Anthropocene," which was held in London, Chicago, Toronto, and Halifax (Canada) during winter 2013/spring 2014. Only after thinking through the discussions and debates elicited from this panel series did it become clear to us that it is not possible to conceive of the Anthropocene independent of history and freedom. This situation sparked an interest in wanting to examine how the very recent recognition of the Anthropocene in just the past decade is connected to the widespread helplessness that comes at the close of the twentieth century regarding the capacity of society to self-consciously transform itself. A.P.M., in particular, owes a debt of gratitude to several pieces published in the *Platypus Review*, most prominently "Decline of the Left in the Twentieth Century" and the "Capital in History: The Need for a Marxian Philosophy of History of the Left," as well as the year-long primary Platypus reading group.

A.M.S. would like to thank the Sociology Graduate Student Association at the University of Kansas—Lawrence, and Shane Wilson, in particular, for the opportunity to present the first version of our argument in an invited talk entitled "Critique of/in the Anthropocene," prepared for the Annual Graduate Student-Sponsored Lecture Series at University of Kansas—Lawrence, February 2014. Earlier versions of our argument were also presented at the

Thirteenth Annual International Social Theory Consortium Conference, University of Tennessee—Knoxville, entitled "Globalization, Critique, and Social Theory: Navigating the Divide between Theory and Practice," Knoxville, Tennessee, May 2014, and at the 109th Annual American Sociological Association Conference, San Francisco, California, August 2014.

We are grateful to Andrew Biro, Harry F. Dahms, and Eric R. Lybeck, whose comments and criticism on earlier versions of this work helped us clarify our ideas further. We thank Laurel Schut for her feedback on the Prologue and Sarah Alford for coming up with the title for the Prologue. A.M.S. would like to thank Harry F. Dahms, in particular, as many of the theoretical ideas developed in this work were first planted in a doctoral dissertation, entitled "Sociobiophysicality, Cold War, and Critical Theory: Human-Ecological Transformation and Contemporary Ecological Subjectivity," authored by A.M.S. (University of Tennessee—Knoxville, 2013) under his supervision.

Photo A (pg. 2) and Photo 1.1 (pg. 31) used with permission courtesy of AKG-images. Photo 2.2 (pg. 67) used with permission courtesy of the Wisconsin Historical Society. We are grateful to NASA for making Photo C (pg. 11), Photo 0.1 (pg. 19), and Photo 2.1 (pg. 63) available under Creative Commons licensing.

palgrave▸pivot

www.palgrave.com/pivot

Prologue: Smog, Haze, Sulfur: The Elusive Clarity of the Anthropocene

Abstract: *This chapter highlights the complex interrelationship between global environmental problems and modern forms of social organization by discussing three examples: (1) the Great London Smog (1952); (2) Southeast Asian Haze (2013); and (3) sulfur in the stratosphere (1815 and the future). In doing so, we illustrate the ways in which seemingly objective laws of modern society undermine efforts to ameliorate societally induced environmental degradation. We conclude by situating the Anthropocene proposal in relation to our current inability to take hold of the runaway character of socioecological development.*

Stoner, Alexander M. and Andony Melathopoulos. *Freedom in the Anthropocene: Twentieth-Century Helplessness in the Face of Climate Change.* New York: Palgrave Macmillan, 2015. DOI: 10.1057/9781137503886.0004.

2   *Freedom in the Anthropocene*

1   The Great London Smog (1952)

PHOTO A   *The Great London Smog*

Source: Street scene with double-decker buses on a smoggy November afternoon in London. This photo was taken in 1953, the year after the deadly smog episode between December 5 and 10, 1952. Photo: AKG-Images.

The air became stagnant as a high-pressure system settled over the Thames River valley on the morning of Friday December 5, 1952. It had been a cold night—another in what had turned out to be an unusually frigid autumn—and in each home a pile of bituminous heating coal smoldered over an open grate. By morning, the city's low-level chimneys had filled the air with smoke. But the air above the city's stagnant hazy curtain warmed more quickly than the air below, resulting in a temperature inversion. Emissions from Europe's largest city were now effectively trapped within a thin band of cold air approximately 100–200 meters deep.

The yellow-gray smog deepened as London's emissions were stoked by the daylight pattern of work and commerce. In the morning people were shuttled to their jobs on the city's new fleet of double-decker diesel buses.[1] Emissions from factories on London's East End kicked in, lending yet another source of coal smoke to the thickening smog. A wide array of public and private activities now worked in concert—from the tail pipes of buses, to the smoke stacks of power-generating stations, and the thousands of stoves preparing Friday evening's dinner—the combined action of a society in motion dropped visibility below 10 meters. By evening, navigating the city became impossible (see Photo A). According to Laskin (2006: 44), "Residents described slow-motion car crashes, pedestrians groping their way along walls and fences, people drowning after stumbling into the River Thames, and long lines of motorists blindly following the taillights of the cars ahead and ending up hopelessly lost in an unusually cool autumn."

Beyond the physical features of their city, the smog obscured the accumulation of deadly levels of fine particulate matter and sulfur dioxide gas. For four days the pollutants were sustained at levels never attained again in any city since. The effect on the lungs of London's war-weakened population was swift, although this was not immediately clear to Londoners. As a physician at a central London hospital recalled, "There was no sense of drama or emergency. It was only when the registrar general published the mortality figures three weeks later that everybody realized that there had, in fact, been a major disaster" (Nagourney, 2003). It is estimated that the Great London Smog claimed the lives of 12,000 people, prompting Hunt and his colleagues (2003: 1209) to declare that "the smog event of 1952 was, in terms of human health effects, the most calamitous of the century."

But preventing the smog by regulating London's pollution was a daunting task: the pollution was not only bound up with how Londoners worked, moved about the city, and generated economic activity, but with their private lives as well. Widely used household heating technology would need to be changed, likely at considerable private expense (Brimblecombe, 2006; Sanderson, 1961). Regulation was additionally problematic given the historic role of the national government in upholding the rights of private householders against the regulatory authority of municipal governments in matters of public health; central regulation would mean remaking long-standing political structures (Keeble, 1978).[2] So it was not altogether surprising when the newly elected Conservative government led by Winston Churchill was reluctant to advance a legislative solution. In fact, the government downplayed the severity of the event (Scarrow, 1972) and seven months passed before they were eventually forced to hold a public enquiry (Keeble, 1978).

Yet, the incident appeared to signal a growing change within society itself. While there had always been discontent with the poor environmental quality of industrial production, it was previously relegated to "a tolerable nuisance" as it appeared interconnected with the capacity of society to generate employment (McNeill, 2000: 59).[3] But a broad and growing discontent with air pollution—one that could not simply be explained as the straightforward reaction to the physical concentration of particulate matter in the atmosphere—began to surface across most industrialized counties after WWII. It was these discontents that became the grounds for the political compact that came to dominate much of the twentieth century, in which the social need for regulating pollution was translated into national environmental legislation enforced and monitored by state agencies. Significantly, the devastation of the Great London Smog positioned the United Kingdom at the forefront of this trend, and in 1956 the House of Commons enacted the Clean Air Act—making the United Kingdom the first country in the world to pass a national statute for regulating air pollution.[4]

Although air quality in the United Kingdom eventually improved, this was largely the product of deep changes taking place in society itself—changes not intended by the regulation enacted under the Clean Air Act. The three decades following WWII (from 1945 to approximately 1973) marked a tremendous transformation, not only in the United Kingdom but across all industrialized countries. The transformation may not have been readily apparent to Londoners the year of the smog as it marked another portentous event: the end of four years of reconstruction aid from the United States via the Marshall Plan. In spite of the aid, the United Kingdom remained highly indebted, and moreover faced considerable turbulence with the double indemnity in 1956 of both a major military defeat in Egypt (the Suez Crisis) and an economic recession (i.e., the same year the Clean Air Act was enacted) (Heyck, 2008). Yet, by the end of the 1950s, the investment in industrial infrastructure was paying dividends, resulting in the most rapid economic expansion seen in the country since the 1870s, topping 4% GDP in 1959 (Heyck, 2008). Moreover, the historically low levels of unemployment that began after the war (i.e., below 2%) were sustained for 30 years, something that had not occurred before or since in the United Kingdom. The persistence of low unemployment was matched by an 80% increase in real earnings across this period, and these factors combined with the increased security created by the newly formed national welfare system, to bring about a sharp rise in domestic consumer spending. By the early 1970s, over half of households owned their own home and these dwellings were well-stocked with newly manufactured appliances such as dishwashers, television sets, and refrigerators (Donnelly, 2014; Heyck, 2008). Over half of the households also owned an automobile.[5]

These social transformations far outpaced the scope of the Clean Air Act and established a pattern whereby attempts to regulate pollution invariably lagged behind changes in society (Brimblecombe, 2006). Coal grates and cooking stoves largely disappeared, not because of new statutes for "smokeless" areas, but because almost a million units of old housing stock were replaced between 1955 and 1970 and switched to inexpensive and more convenient all-electric or gas-fired appliances (Scarrow, 1972). Industrial coal smoke also left the inner city as heavy industry was largely phased out and replaced with new and more efficient industries centered on the manufacture of automotives, electronics, and aerospace in factories located in new suburban developments. Moreover, coal use itself began to decline. A month after the passage of the Clean Air Act, the world's first nuclear power-generating plant was connected to the British electrical grid. Flash forward to a decade later and domestic coal consumption had declined by almost 40% and was set to decline even more precipitously with the discovery of natural gas in the North Sea in 1967. It has now been almost 40 years since Britain built its last coal-fired plant.

Yet, the history of pollution from 1952 to the present is not one of straightforward social progress. It might be more accurate to say that the problem of atmospheric pollution has not so much been improved but *transformed*. Although coal smoke and sulfur dioxide have grown to be relatively minor problems in the United Kingdom, as it has across all the industrial centers that emerged in the decades following WWII (e.g., in London, Tokyo, Rhine-Ruhr Germany, the United States Rust Belt), pollution from coal burning has shifted to new centers such as Shanghai (China), Ho Chi Minh City (Vietnam), and Sao Paulo (Brazil) (Fenger, 1999). However, the decline of public transit relative to the rise in automobile traffic has generated new pollutants in cities like London, including volatile organic compounds, nitrogen oxides, and ozone (Brimblecombe, 2006). More significantly, pollution has ceased to be something that only causes harm in the region adjacent to the emission source, as the scale at which the effects of pollution are felt has telescoped. Today, the most mundane features of social life (such as driving a car to work) effectively contribute to the alteration of the planetary systems that regulate climate. The form of society that ultimately raised the Great Smog of London in 1952 continues to exceed efforts to meaningfully regulate it.

2 Southeastern Asian Haze (2013)

Unlike the Great London Smog, the Southeastern Asian Haze of 2013 did not originate in the two hardest hit cities: Singapore and Kuala Lumpur. Instead, it was carried by winds across the narrow Strait of Malacca from massive fires[6] hundreds of kilometers to the west in the Riau province in eastern Sumatra, Indonesia, as shown in Photo B. Levels of particulate pollution in Singapore (the worst hit of the two cities) doubled previous records, as the Pollutant Standard Index reached levels deemed "very unhealthy" for three consecutive days (Gaveau et al., 2014). Although the fires burned for a little less than two weeks (the bulk of the burning was confined to a five-day span), they constituted 5–10% of all Indonesian greenhouse gas (GHG) emissions in 2013 (Gaveau et al., 2014). These unusually high emissions, however, did not result from the burning of the forest, but from the underlying tropical peat—the largest near-surface reserve of terrestrial carbon on the planet. Indeed, there is more carbon sequestered in the region's tropical peat than all the proven reserves of oil on the planet.[7]

The Great London Smog and the Southeast Asian Haze bracket the era of environmental politics in a remarkable way. If the Great London Smog expressed the need for society to better regulate environmental quality, the Southeast Asian Haze represents a dimension of how society subsequently took up that task. Although the fires were plainly associated with

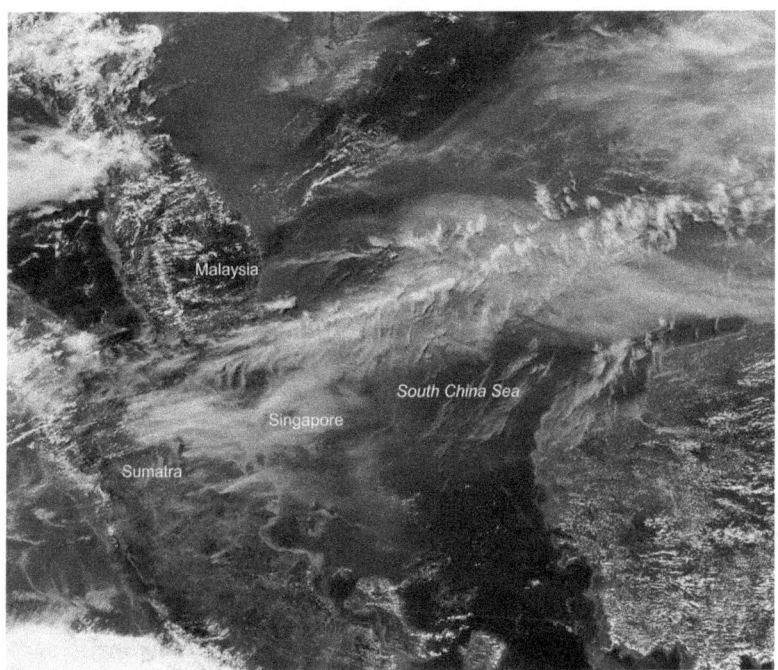

PHOTO B *Southeastern Asian Haze*

Source: A satellite image taken on June 19, 2013, of smoke from wildfires from oil palm-growing regions on the Indonesian island of Sumatra blowing east toward southern Malaysia and Singapore. The smoke formed a thick haze in Singapore and by the evening broke all previous record for smog severity. But even these levels were exceeded in subsequent days. Photo: NASA Earth Observatory.

clearing land for the expansion of oil palm plantations (Gaveau et al., 2014), this expansion was, ironically, intended not to increase, but rather to decrease GHG emissions.[8]

Besides being used as a feedstock in the manufacture of a vast array of products (from lipstick to chocolate), oil palm has become a key part of the EU's renewable energy strategy through its role in biodiesel production (Field et al., 2008; Koh and Ghazoul, 2008). In principle, blending biofuels with fossil fuels slows GHG emissions because biofuel feedstock (e.g., oil palm, but also soybean and temperate zone oilseed rape) fixes atmospheric carbon dioxide as it grows. In the early 2000s this principle was met with enthusiastic support from policy-makers, particularly in the United States and the Europe Union, who were concerned with unrelated but persistent problems—namely, low farm income and the internal political pressure exerted by national farm lobby groups (Potter, 2009).

The threat of low agricultural commodity prices has plagued the farm sector through most of the twentieth century because agricultural output has tended to far surpass the amount of food that could be purchased (Friedmann, 1982; 1993). The chronic tendency of agriculture toward oversupply persisted even into the period of accelerating consumer affluence (1945–1973) and has only episodically experienced periods of high profits (e.g., 1973–1980 and 2006–2011). The first of these crises of farm income began with the collapse of the grain market in 1921 and extended well into the 1930s (Friedmann, 1982; 1993). The collapse of agricultural prices constituted a broad social crisis, since the majority of people in industrialized countries before WWII lived in rural areas. In response, national governments in the 1930s began managing agricultural prices through various forms of subsidization (e.g., import tariffs, supply management programs, payments for removing land from production, and direct payments) (Friedmann, 1982; 1993). But farm income support programs were themselves drawn into a deep crisis in the mid-1970s along with many other features of the post-WWII political and economic order.[9] Although the crisis resulted in a dramatic spike in the prices of all agricultural commodities, these increases quickly evaporated into two decades of chronic low profits.[10] As governments in these countries attempted to restructure their economies by reducing government expenditures and promoting global trade liberalization, farm income programs became increasingly difficult to justify to a primarily urbanized tax base already facing deep cuts to their own state welfare benefits (Potter, 1998; 2009).

Significantly, the critique of intensive farm management practices was one of the foundations of the modern environmental movement.[11] At the peak of the farm crisis in the 1990s, environmental groups called for state subsidies to be conditional on farmers using less intensive farm practices (Cain and Lovejoy, 2004; McGranahan et al., 2013; Potter, 1998; 2009). And state-subsidized biofuel initiatives provided an attractive mechanism to connect the concerns of urban environmentalists with rural farm communities. By producing biofuels from crops such as corn (biogasoline) or oilseed rape and soybean (biodiesel), stocks of these crops could be drawn down (increasing prices), while their displacement of fossil fuels could simultaneously reduce overall GHG emissions.

Consequently, both the United States (2002) and EU member states (2003) aggressively legislated biofuel blending standards (Gerasimchuk and Koh, 2013; Tyner, 2008).[12]

Blending standards initially increased demand for EU oilseed crops, but the excess supply of these crops was almost completely absorbed by 2006, necessitating the import of oil palm to make up the shortfall.[13] On the eve of the Southeastern Asian Haze, oil palm constituted 20% of the feedstock for EU biodiesel (Gerasimchuk and Koh, 2013). While the decades long problem of farm income was solved by a biofuel strategy, with food prices surging between 2006 and 2011 (Headey and Fan, 2008), it also led to the aggressive expansion of oil palm, which, paradoxically, hastened carbon emissions through the destruction of peat forests (Koh and Ghazoul, 2010). In light of the evidence that biofuels made from crop plants (first-generation biofuels) do not reduce GHG reduction emissions, environmental groups have increasingly applied pressure on legislators to mandate second-generation biofuels (made from noncrop plant sources) which would significantly reduce the farm price support dimensions of the initial strategy.[14] Sensing an opportunity, as well as a crumbling coalition, farm groups have responded by arguing that recent high prices prove the need for national self-sufficiency in food production and have called for a return to direct support of EU farmers (Potter, 2009). In fact, while EU environmentalist and farm groups find themselves increasingly on opposing sides of biofuel policy, they find common cause in the protectionist demand to limit the importation of oil palm from Indonesia and Malaysia. Yet, while such a policy may satisfy the interests of EU oilseed producers, it does so at the expense of the oil palm sector, which employs millions of people in Malaysia and Indonesia (Mukherjee and Sovacool, 2014).

The 2013 Southeast Asia Haze highlights how seemingly objective laws of modern society (e.g., the chronic tendency of agriculture toward oversupply) invariably limit our capacity to regulate damaging forms of pollution. Although these "laws" emerge from society itself, they appear rigid and unchangeable. Yet, while some astute and powerfully positioned actors can certainly advance their interests over others (e.g., EU farmers over Southeast Asian oil palm companies), all remain subject to a social context that no one controls. Moreover, our ability to provide lasting solutions to historic problems, such as farm income, not only seem increasingly beyond our reach but tend to compound with other unresolved problems, conditioning our ability

3 Sulfur and the stratosphere (1815 and the future)

> I had a dream, which was not all a dream.
> The bright sun was extinguish'd, and the stars
> Did wander darkling in the eternal space,
> Rayless, and pathless, and the icy earth
> Swung blind and blackening in the moonless air;
> Morn came and went – and came, and brought no day,
> And men forgot their passions in the dread
> Of this their desolation; and all hearts
> Were chill'd into a selfish prayer for light
>
> — Excerpt from *Darkness,* Lord Byron, 1816

The largest volcanic eruption in recorded history began on the evening of Wednesday April 5, 1815, on Mount Tambora on the small Indonesian island of Sumbawa, 300 km east of Bali (Oppenheimer, 2003). The first eruption was massive in its own right and was mistaken for canon fires some 350 km northeast in Sulawesi. The second eruption, however, has no parallel in recorded history (Photo C).[15] The explosions began early in the evening on Monday April 10 and over the next four days 50 cubic kilometers of material flowed out as ash and magma, equivalent to 2,000 times the volume of concrete in China's massive Three Gorges hydroelectric dam. An immense amount of ash buried crops over an area four times the size of the United Kingdom and darkened the sky for two days. A tsunami of 1–4 m was cast out and stretched across a range of at least 1,200 km. Lava flows extended 20 km out from the mountain, filling the island with toxic gasses that poisoned many residents. The local impact of the eruption was, without a doubt, the most devastating in history with an estimated 71,000 dead (Oppenheimer, 2003).

The devastation of the eruption, however, was by no means restricted to the Indonesian archipelago. The plume of ash and gas rocketed over 40 km into the sky, well above the troposphere (the band containing the earth's weather and clouds) and into the stratosphere. It is estimated that a 60 teragrams of sulfur (ten times that of the Mount Pinatubo (Philippines) eruption in June 1991) reached the stratosphere where the small particles were carried by winds longitudinally across the planet,

PHOTO C *Volcanoes and the stratosphere*

Source: Alaska's Pavlof volcano eruption on May 18, 2013. Small eruptions like these are common along the Aleutian Islands, involve the displacement of less than 0.100 cubic kilometers of dense rock, and generate columns of smoke less than 15 kilometers in height. Much less common are the supereruptions like that of Mount Tambora on April 10, 1815, which involve the displacement of over 300 cubic kilometers of rock and an ash cloud extending 40 kilometers (i.e., well into the stratosphere). The cooling effect of stratospheric ash from Tambora resulted in dramatic cooling in Northern Europe and the Northeastern United States the following summer. Climate scientists are presently looking to mimic this cooling effect as way to mitigate climate change by developing Solar-Radiation Management (SRM) technology. Photo: The Alaska Volcano Observatory/Rob Gutro, NASA Goddard Space Flight Center.

eventually concentrating at latitudes equivalent to Northern Europe and the northern states of the United States. The particles lingered for up to three years before falling back to earth as acidified precipitation (Stothers, 1984).

The particle clouds were first observed over England some three months after the eruption and reached peak intensity by early autumn. They appeared as richly colored sunsets and twilight during the last week of June 1815.[16] As the dust continued to mix, it gave the illusion that stars in the night sky trembled. The darkening stratosphere meant sunspots became visible.

The spectacular atmospheric display of 1815, however, was a harbinger of an oncoming disaster. The particles absorbed and reflected back solar radiation, leading to three years of historically unprecedented cold weather. In fact, 1816 became known as the "year without summer" as temperatures in Europe fell to the lowest they had been since the deepest years of the Little Ice Age in the early seventeenth century. The cool wet summers extended across Europe's grain-growing regions, resulting in a trebling of prices and shortages of bread. Famine extended across northeastern France, southwestern Germany, Italy, Ireland, eastern Switzerland, and across the Hapsburg Empire, surpassing even the worst famines of the eighteenth century (Post, 1977). The cold extended to North America where repeated and extensive crop failures in the New England region of the United States prompted widespread migration to Ohio.

Coming in the wake of 25 years of continuous war and the decisive defeat of Napoleon in July 1815,[17] Byron's 1816 poem *Darkness* reflects the sense of doom that shrouded Europe as the cold set in. The inner crisis of society following the French Revolution appeared to find expression in the sky; the dramatic shift of sky and climate sounded the tocsin of social decline. The sense of progress that marked the eighteenth-century Enlightenment was enveloped in Romantic recoil. In the wake of the French Revolution and the 1815 restoration of the old order at the Congress of Vienna, an antiliberal political reaction appeared to be taking grip of the continent. But change could not be so easily contained.

As the sulfur fell back to the earth (1819), Benjamin Constant would address the Athénée Royal in Paris and declare that in spite of the restoration of the monarchy in France, society could not be rolled back into its old form.[18] A new form of society, one without historical precedent, was being articulated. Both the 1952 London Smog and the 2013 Southeast Asian Haze attest to the legacy of the relentless and unimaginable social and biophysical transformations that continue to advance. What Byron's despair failed to recognize, however, was the accomplishment of *bourgeois society*, in which the productivity of humanity had fully moved from the rural forms of peasant production to production in cities—and the rise of the Third Estate—the class of "commoners" who were to be judged not on the basis of tradition or divine orders but on their capacity to "work." Taking the course of humanity out of the hands of "fate," in a sense denaturalizing it, and drawing it into the sphere of responsible human activity was bound up with the "freedom thinking" of figures such as Jean Jacques Rousseau, Immanuel Kant, Adam Smith, and Georg Friedrich Hegel and the attempt to advance actual freedom arising in the politics of what become known as the Left.

Yet, the pattern of progress throughout the nineteenth and twentieth centuries has not been one of society freely determining itself. Both the 1952 Great London Smog and the 2013 Southeast Asian Haze point toward deep social structures that have a motion of their own and to which we are all ultimately subject. The question of how society might place itself at the service of freedom has progressively grown dim. Today, society threatens to transform planetary systems in a manner that surpasses the reach of geological processes of much larger timescales. Taking up the task of improving environmental quality means taking hold of broad social processes that are seemingly more complex and incomprehensible than the dynamic chemical processes and patterns of aerosol movements in the stratosphere. Consequently, tackling a problem such as climate change (e.g., by charting a path to renewable fuels) seems to only get mired in long-standing social–structural problems (e.g., agricultural overproduction).

At the same time, the obstacles associated with reducing GHG emissions have given rise to an array of increasingly sophisticated technical solutions to climate change. One of the most controversial has been Solar-Radiation Management (SRM) technologies designed to master the process of planetary cooling that follow volcanic eruptions such as Tambora. The prominent Nobel Prize-winning atmospheric chemist Paul Crutzen explained the growing need for SRM technology in a recent editorial for the journal *Climate Change* (2006: 211–212):

> By far the preferred way... is to lower the emissions of the greenhouse gases. However, so far, attempts in that direction have been grossly unsuccessful. While stabilization of CO_2 would require a 60–80% reduction in current anthropogenic CO_2 emissions, worldwide they actually increased by 2% from 2001 to 2002, a trend, which probably will not change at least for the remaining 6-year term of the Kyoto protocol, further increasing the required emission restrictions. Therefore, although by far not the best solution, the usefulness of artificially enhancing earth's albedo and thereby cooling climate by adding sunlight reflecting aerosol in the stratosphere might again be explored and debated as a way to defuse the Catch-22 situation... [and] counteract the climate forcing of growing CO_2 emissions.

Significantly, Crutzen's response indicates the extent to which SRM technology emerges from the persisting failure to regulate pollution at the level of society. Although the very possibility of something like SRM technology would have struck figures in the Enlightenment as a great accomplishment, today it has become an index of the inability of society to freely regulate itself. Climate and atmospheric scientists have grown frustrated by efforts to reduce climate change "from below" (i.e., at the level of civil society) and have resigned themselves to technical solutions to resolve the impasse "from above." But even here the scientists find themselves increasingly blocked by the political inertia around approving small-scale trials to better understand and model the dynamics and risks of stratospheric sulfur deposition (Keith et al., 2014). Yet inaction on SRM research, in the face of accelerating climate change, may result in even deeper dilemmas: "it would be reckless to conduct the first large-scale SRM in an emergency" (Keith et al., 2010: 426).

Crutzen and his colleagues have proposed the Anthropocene to demarcate the new geological epoch that came into being around the time of the Tambora eruption in 1815.[19] Looking back over the past two centuries we can discern something of this period by following a single element—sulfur—across history. Sulfur attains a new planetary significance through the growing crisis of human society. In 1952, there was a sense that that sulfur dioxide was an object to be shaped by government regulation. Yet this regulation was almost immediately outpaced by the explosive changes that took place in society after WWII. While the form of pollution changed, new unanticipated ecological problems emerged. But these new problems appear even more deeply connected to long-standing problems within society (and hence seem more intractable).

We now face ballooning ecological degradation that pales in comparison not only to past environmental problems (e.g., 1952), but even the kind of biophysical transformations posed by episodic geological processes (e.g., 1815). Today, sulfur presents itself to us not as a constituent part of the planet to be regulated by society—as an object through which the potential of humanity could be realized—but as an inevitability, as something that we will collectively be forced to shoot into the stratosphere because of our inability to take hold of the runaway character of society itself.

Notes

1 The program to replace the aging electric tram system with the iconic red buses had just been completed five months earlier.

2 The 1936 Public Health Act, for example, placed restrictions on municipal governments from being able to prosecute homeowners for issues associated with domestic chimneys (Keeble, 1978: 262).
3 As McNeill (2000: 93) recounts, the nineteenth-century civic anthem of Japan's first large steel town, the advancement of coal-based industry was viewed as a source of civic pride: "Billows of smoke filling the sky. Our steel plant, a grandeur unmatched: Yawata, O Yawata, our city!"
4 Significantly, the 1956 Clean Air Act predated even the landmark US 1970 Clean Air Act by more than a decade. A decisive factor in the passing of the Act in the United Kingdom was the emergence of the National Smoke Abatement Society, a relatively obscure lobbying group at the time. The Society grew to prominence after the smog and its concerted lobbying effort played a pivotal role in advancing the legislation (Scarrow, 1972). Extra-parliamentary environmental interest groups subsequently became a key element in the formation of national environmental policy in other industrial countries. Paradoxically, in spite of the success of the Society in advancing the 1956 Clean Air Act, the role of environmental groups in the United Kingdom remained quite small until the 1990s (McGrew, 1990) and was dwarfed by the broad movements that emerged in the 1970s in both the United States (Gottlieb, 2005 [1994]) and West Germany (Mewes, 1983).
5 But, as we discuss in Chapter 3, the transformations expanded far beyond economic dimensions, across to almost every facet of life. The distinct class differences in British society in the nineteenth century seemed to erode as patterns of fashion and culture changed into the "classless" world of blue jeans and the pop music (Donnelly, 2014). In their place social discontents increasingly centered on new divisions—race and gender—as low unemployment resulted in the massive influx of immigrants from the West Indies, Africa, India, and Pakistan, and enabled women to enter the workforce in large numbers (Heyck, 2008). Moreover, sustained affluence paradoxically generated a counterculture that questioned the desirability of the current and prosperous form of society, which significantly included the environmental movement (McGrew, 1990).
6 The fires covered 163,000 hectares, just over half of the area of Yosemite National Park in California.
7 But this was by no means the first such fire. A much larger Indonesian fire in 1997 released anywhere from 13 to 40% of the annual carbon generated from the combustion of fossil fuels *globally* that year (Page et al., 2002). In fact, the 1997 fire resulted in the largest annual increase in atmospheric carbon dioxide ever recorded.
8 There is little doubt that the fires are connected to the expansion of oil palm. Careful analysis of 2013 LANDSAT satellite imagery highlights a peculiar socioecological dynamic set off by the increased demand for biodiesel.

The expansion of oil palm has largely come at the expense of the regions' forest cover (Koh and Ghazoul, 2008; Koh and Wilcove, 2007). Reduced forest cover means even relatively short dry periods (less than 2 months) increase fire hazard significantly (e.g., by increasing soil temperature and changing hydrological patterns). Consequently, blazes can occur even in a wetter-than-average year like 2013 (Gaveau et al., 2014). This changed fire ecology combined with a dispersed pattern of ignition coinciding with the spatial arrangement of oil palm producers and state land use zoning. Conspicuously, over 80% of the burned land was recently cleared of forest and had either been allocated by the state to oil palm companies or had contested ownership. Fire—the quickest and most cost-effective manner of clearing coarse woody debris from such fields—was being targeted to land most suitable for expanding oil palm. Within a month aerial drones spotted excavators preparing the recently burned site for planting. Biophysical dimensions of the forest, including fire ecology, hydrological cycles, not to mention planetary carbon regulation, were interacting with multiple dimensions of society. These social dimensions include not only the oil palm producers, who are driven by the stable income stream associated with oil palm and the growing awareness of global environmental threats that initially motivated biofuel policy but other seemingly unrelated sociohistorical pressures as well.

9   The long period of social stability in industrial countries following WWII came to an abrupt end at the beginning of the 1970s. This crisis had multiple dimensions and was felt internationally, but a key feature took the form of a severe economic crisis in the United States, the key economic engine of the post-WWII reconstruction. A mixture of high inflation and persistent unemployment (stagflation) took hold at the end of the 1960s prompting the Nixon government to implement wage and price controls in 1971. These economic problems were immediately compounded by a steep rise in oil prices following an energy embargo by the Organization of Petroleum Exporting Countries (OPEC) in 1973. The economic slowdown strongly conditioned the restructuring of farm income support programs, particularly in the 1985 US Farm Bill (McGranahan et al., 2013; Potter, 1998).

10 The sag in the farm sector was initiated by a crash in farm commodity prices in the 1980s followed by crippling double-digit interest rates (Potter, 1998). By the end of the 1980s, over 70% of large farms in the United States were at risk of liquidation and expenditure on commodity support programs soared to $26 billion. Similarly, in the EU, 70% of the entire commission budget was devoted to some form of farm income support (Friedmann, 1982; 1993).

11 Racheal Carson's book *Silent Spring* (1962) is frequently credited with the birth of the modern environmental movement. The book focused on the

negative side effects of using pesticides such as DDT to manage insect pests in intensive agricultural systems.

12 As a result, all EU diesel now contains at least 5.75% biodiesel and is set to rise to 10% in 2020.

13 The amount of crops required to meet the blending standards proved immense. The carbon currently emitted through the combustion of fossil fuels roughly equals the amount of carbon fixed by all agricultural crops on the planet (Field et al., 2008). Moreover, the demand for liquid transport fuels is accelerating, such that by 2050 an anticipated of 2.26 billion new cars would require an estimated 54 million hectares dedicated to growing biofuel crops (Ghazoul et al., 2010).

14 There are attempts underway in the EU to renegotiate the renewable fuel directive to better address the issue of GHG emissions by adding so-called Indirect Land Use Change (ILUC) provisions that mandate the amount of second-generation noncrop sources of biofuel do not change agricultural land use patterns (e.g., from algae, waste, and organic cellulosic residues). Predictably, these new provisions are being vigorously opposed by farm groups.

15 The Tambora eruption is regarded as more massive in scale than even the Aegean Minoan eruption 3,600 years earlier, and only exceeded eruption from the Campi Flegrei caldera in Italy 40,000 years ago (Oppenheimer, 2003: 253).

16 The sky had a characteristic look: orange-red horizon crested with deep purples and pinks that occasionally streaked with dark bands. These were exemplified in the work of that period from the British Romantic painter J.M.W. Turner in pieces such as "Sunset" (c. 1833) (Zerefos et al., 2007).

17 The Tambora eruption occurred just as the Great Powers of Europe (the Austrian (Habsburg) Empire, France, Prussia, Russia, and the British Empire) began mobilizing their armies for the decisive defeat of Napoleon in July 1815.

18 As Constant will argue this new form of society is characterized by a fundamentally new form of political liberty that emerges from free individuals engaged in commerce: "Commerce makes the action of arbitrary power over our existence more oppressive than in the past, because, as our speculations are more varied, arbitrary power must multiply itself to reach them. But commerce also makes the action of arbitrary power easier to elude, because it changes the nature of property, which becomes, in virtue of this change, almost impossible to seize... circulation creates an invisible and invincible obstacle to the actions of social power" (Constant, 1988 [1819]: 324–325).

19 The Tambora eruption can readily be discerned in ice cores and has been proposed as a marker for the onset of the Anthropocene (Smith, 2014).

DOI: 10.1057/9781137503886.0004

Introduction: What Is the Meaning of Freedom in the Anthropocene?

Abstract: *This chapter summarizes the Anthropocene periodization, as proposed by Paul Crutzen and his colleagues. After reviewing the literature, we advance the proposition that the Anthropocene is unable to address the problem of freedom implicated in the environment–society problematic—a seemingly paradoxical process wherein environmental degradation increases amid the growth of environmental attention and concern—because it is unclear of its causes and how one might plausibly overcome this context while being part of it. We then outline the main argument of the book. And finally, we conclude by providing a breakdown of subsequent chapters.*

Stoner, Alexander M. and Andony Melathopoulos. *Freedom in the Anthropocene: Twentieth-Century Helplessness in the Face of Climate Change.* New York: Palgrave Macmillan, 2015. DOI: 10.1057/9781137503886.0005.

Introduction 19

PHOTO O.1 *The Anthropocene*

Source: An evening photograph from the International Space Station of the Atlantic coast of the United States (February 6, 2012). The large interconnected metropolitan areas clearly visible from light emissions show the expanse of human development that characterizes the Anthropocene. Photo: NASA.

The Anthropocene distinguishes a new geological epoch in which the activity of human societies has become the determinant force in the changing structure of the Earth's planetary systems (reviewed in Ruddiman, 2014). The Anthropocene demarcates the period from the Industrial Revolution in the eighteenth century to the present as capturing a change in the history of the planet. Although the Anthropocene is not a formally defined geological unit within the Geological Time Scale, a proposal for its inclusion is being developed and will be considered by the International Commission on Stratigraphy by 2016.[1]

While we are ostensibly 250 years into the Anthropocene, our recognition of it comes with a conspicuous lag; James Watt and other members of the Scottish Enlightenment certainly would not have recognized themselves as living at the dawn of the Anthropocene. The idea of a Holocene/Anthropocene boundary, in fact, has only gained a following among environmental scientists, sociologists and the broader public in the last decade.[2] Why the delay? The first impulse is to ascribe such late recognition to a technical problem, namely that the complex models and instrumentation necessary to register changes in planetary systems were only developed in the last 40 years (Steffen et al., 2011). Another is to point out that biophysical impact becomes more acute and noticeable as time passes, particularly following the sharp increase in consumption and global interconnectivity that started in the years following WWII, a phase of the Anthropocene termed the *Great Acceleration* (Crutzen and Stoermer, 2000). We only now see this period as the Anthropocene because its ecological implications have only now become apparent.

In this book we will argue for an additional reason for the very recent recognition of the Anthropocene. This has to do with the growing doubts that come at the close of the twentieth century regarding the capacity of society to self-consciously transform itself. While a person like James Watt would have held the aspirations for the free development and transformation of society from the constraints of feudalism—that is, the idea and political project for freedom—our moment is marked by a dramatic attenuation, or even distrust, that such transformation is even desirable (Latour, 2014). The arrival of the Anthropocene coincides conspicuously with a sense that the twentieth century was characterized not by freedom, but the wholesale return of structural constraints that restrict self-conscious social transformation, bringing with it a profound sense of helplessness (Postone, 2006). The specific interest of environmental scientists and sociologists in history, we contend, is bound up with the fact that it is precisely history that appears to ensnare human society in a runaway developmental pattern that will not lead to the opening of human capacities and the flourishing of ecosystems, but rather to the inevitable diminishing of both.

For this reason, the Anthropocene may not have even been plausible to those making the very recent transition into the Great Acceleration. The Great Acceleration, in fact, not only delimits a period of compounding human impact on the environment, but importantly, one in which public concern for and attention to this impact broadly emerged. Significantly, however, the Great Acceleration is *not* the history of how increasing awareness of human impact on the environment translated into our capacity to ameliorate societally induced environmental degradation. Paradoxically, environmental degradation compounded in proportion to our awareness of these problems (Stoner, 2014). While the growth of environmental movements in North America and Europe in the 1970s and early 1980s appeared to signal the emergence of historically new political forms, these found themselves mired in a largely rearguard battle. From the "implementation deficit" that followed the creation of national government environmental agencies in the early 1970s in many OECD countries (Røpke, 2005: 266), to the failure to guide economic development along the sustainable development framework outlined by the Brundtland Report in 1987 (Blühdorn, 2007; Steffen, 2008), to the sense of helplessness that attends the most recent turn to protest-oriented climate justice (Blühdorn, 2013; Karlsson, 2013), the understated feature of the Great Acceleration is the extent to which the runaway development pattern seems foreclosed in spite of numerous attempts to change its course.

The growing gap between the broad awareness of environmental problems, the subjective dimension of societally induced environmental degradation (Stoner, 2014), and our ability to transform the objective dimension of the world in accordance with this awareness is of growing concern to environmental scientists and sociologists (Balmford and Cowling, 2006; Fischer et al., 2007; Steffen et al., 2008; 2012). Just as their disciplines have matured, and can therefore (presumably) bring their accumulated weight of theoretical and empirical knowledge to bear across innumerable conservation initiatives, their actions in the world end up being frustrated and turning into "politically mediated compromises that fall far short" (Fischer et al., 2007: 623). We contend that the current interest in the Anthropocene is connected to this sense of helplessness, but in a way that never fully confronts its paradoxical character.

While the Anthropocene characterizes the global significance of human–ecological transformation as having relatively recent origins (the beginning of the Industrial Revolution in Europe) and although it demarcates transitions within the Anthropocene (e.g., the Great Acceleration), we contend that its account of what generates change ends up linear and without the potential for being otherwise. This failure to specify the type of transformation and its conditions of possibility ends up in the externalization of the factors of transformation. Consequently, scholars have yet to specify the dynamic that is generative of the current environment–society crisis wherein rising consciousness of environmental degradation does not translate into action.[3] What remains paradoxical is the fact that "the intensity and scale of societally induced environmental degradation, which rose to historically unprecedented levels during the latter half of the twentieth century, is synchronous with an equally impressive increase in public concern for and attention to the well-being of the biophysical world" (Stoner, 2014: 622). In this book we will refer to this paradox (wherein environmental degradation increases amid the growth of environmental attention and concern) as the *environment–society problematic*.

Examining the environment–society problematic, we further distinguish between an "objective" dimension (actual, concrete human–ecological transformation) and a "subjective" dimension (the social conception and understanding of the natural environment). Objective human–ecological transformation is a property of human labor. Human labor is a basic mediation between humans and their environment and the ways in which, through such laboring activity, both humans and environment are transformed in the process of meeting a given end. In modern capitalist society, however, human labor takes a particular form, which Marx (1988 [1844]) termed *alienation*—a dynamic process, constituted by the capitalist mode of production through which humans are estranged from self, nature, others, and consciousness in such a way so as to inhibit these very same humans from consciously recognizing that this is indeed the case (Stoner, 2014: 632). Following Stoner (2014), the underlying dynamics fueling the environment–society problematic must be understood theoretically with reference to alienation: alienation mediates the reciprocity between the objective and the subjective dimensions of the environment–society problematic. The reader should bear in mind that in examining the Anthropocene we are presupposing a historical context (modern capitalist society) constituted by alienated social relations. In this context, rather than self-consciously transforming the socio-biophysical world, people are dominated and controlled by the history they create. What this means will become clearer in the chapters that follow.

The subjective dimension of the environment–society problematic, like its objective dimension, is also historically specific. As we elaborate in Chapter 3, a particular social conception and understanding of the natural environment, which encompasses a wide variety of contemporary ecological thought, emerged throughout the latter half of the twentieth century. This contemporary ecological subjectivity can only be adequately understood in connection to objective social structure, including large-scale structural transformations of global capitalism. Although inextricably intertwined, the subjective–objective dimensions (i.e., consciousness and social structure) of the environment–society problematic are nonidentical (Stoner, 2014). As we shall see, one of the peculiar features of the present historical moment, which has a significant bearing on contemporary ecological subjectivity, is an apparent decontextualization in which subject–object dimensions are conflated and, as such, appear identical. Consequently, contemporary ecological subjectivity is marked by the appearance of a direct causal relationship between societally induced environmental degradation, on the one hand, and widespread growth of environmental attention and concern, on the other. The issue here, which we shall return to elaborate throughout the course of this book, revolves around the failure to recognize the material contradictions currently inhibiting growing efforts to ameliorate societally induced environmental degradation.

What is paradoxical about the environment–society problematic is that the expansion of ecological consciousness has not yet translated into revolutionary transformation of society and culture worldwide in the face of the objective imperative to do so. Admittedly, one could suggest that this is not at all paradoxical. And one could provide all sorts of explanations for why this is the case, including, for example, vested "interests," denial, cynicism, and so on. However, what we are interested in, and what we hope to render plausible throughout the course of this book, is that while ecological degradation is becoming increasingly visible and less deniable, the paradoxical process at work remains largely concealed. As it turns out, things are getting worse on the road to catastrophe, both in the objective sense of worsening harm to increasing numbers of living beings, human and otherwise, and in the subjective sense, which masks the fact that nothing new under the sun ails us but the alienated core of our modern being.

In this book we will argue that although the Anthropocene is an attempt to understand history, it does so in a way that is unable to address the problem of freedom implicated in the environment–society problematic because it is unclear of its causes and how one can overcome this context while being part of it. As a result, the Anthropocene avoids the central questions surrounding the problem of freedom: How has history qualitatively changed in the last 200 years? What is the connection between subjectivity and social–structural transformation, how has this problem shifted in the twentieth century? And why, from the standpoint of the present, does society appear to be constituted by a "runaway" alien development pattern that is the basis of our current ecological predicament?

We employ critical theory to uncover how the Anthropocene projects our helplessness in the present backward over all of history, thereby naturalizing what is historically specific about the relationship between modern society and the natural environment. Rather than dismissing the concept of the Anthropocene out-of-hand, Chapters 1–3 review the last century of the Anthropocene through the lens of three twentieth-century critical theorists: Georg Lukács (1885–1971), Theodor W. Adorno (1903–1969), and Moishe Postone (1942–), and their critiques of reification, identity thinking, and traditional Marxism, respectively. Given that Lukács, Adorno, and Postone each approached the problem of history and freedom in their moment, coinciding, respectively, with the emergence (1920s), development (1960s), and present moment of the Great Acceleration, the meaning of freedom in the Anthropocene is clarified by reinterpreting the core critiques these theorists engaged. This schema is depicted visually in Figure 0.1.

Key events raised in the book marking three periods of the Anthropocene outlined in Steffen et al. (2007) against atmospheric carbon dioxide (CO_2) record from ice core data before 1958 (Etheridge et al., 1996; MacFarling Meure et al., 2006) and the yearly averages of direct observations from Mauna Loa and the South Pole after and including 1958 from Scripps CO_2 Program (http://scrippsco2.ucsd.edu). Key events in the book are overlaid on the graph, as well as publication dates of the three critiques raised in Chapters 1–3.

While Crutzen and his colleagues largely ascribe the emergence of the Industrial Revolution to factors exogenous to society, namely the

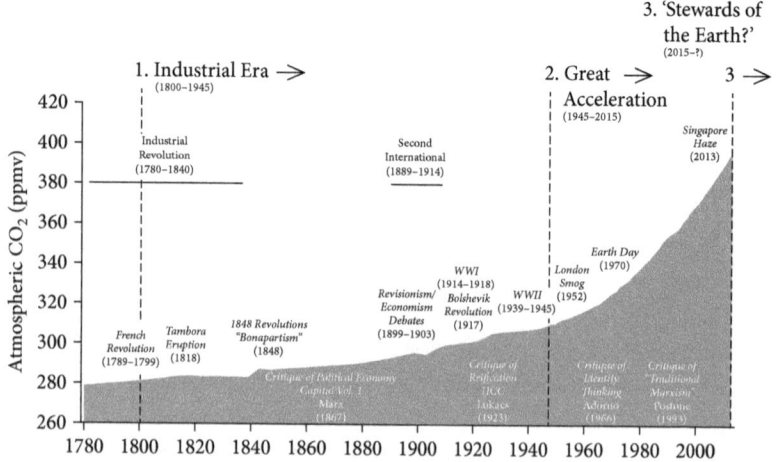

FIGURE 0.1 *Freedom and critique in the Anthropocene*

availability of fossil fuels and the machines they power (Steffen et al., 2007; 2011), Lukács, Adorno, and Postone each attempt to understand the changes taking place in their moment as linked to a social dynamic that coincides with the dawn of the Anthropocene, namely the emergence of the capitalist market economy and the development of historically unprecedented ways of organizing human societies around free wage labor. In doing so, these theorists work through the implications of Karl Marx's (1818–1883) critical theory of modern society under different historical circumstances. In brief, these circumstances were the crisis of the revolutionary Marxist-led Second Socialist International of the early twentieth century (Lukács); the clear failure of these politics to lead to self-conscious social transformation by the 1930s, the rise of authoritarianism, and the concomitant consolidation of social domination in capital (Adorno); and finally, the reversal of state capitalism after 1973, followed by the global penetration of neoliberal capitalism without the attendant growth of internationalist Marxian mass politics (Postone).

Chapter 1 analyzes Georg Lukács' critique of reification as elaborated in his (1923) collection of essays, *History and Class Consciousness*, in order to glean insight into the problem of history and freedom during Lukács' moment, which coincides with the emergence of the Great Acceleration.[4] The chapter begins by situating Lukács' critique of reification against the background of the crisis of the revolutionary Marxist-led Second Socialist International. We focus specifically on the debates, largely among the Marxist leadership, over the question of interpreting the content of the relationship between socialism, the workers movement, and the new and emerging forms of the state. Amid these developments, Lukács' critique of reification redirects attention to the subjective quality of worker struggles and how these struggles could still exert social transformation, but in a way that gets swept along with historical currents that increasingly take the appearance of being natural (i.e., "reified"). Since this distinctive phase of capitalist development appears to anticipate the pattern of state capitalism that gave the Great Acceleration its form, Lukács' critique of reification is particularly relevant to the last century of the Anthropocene.

Chapter 2 engages Theodore W. Adorno's critique of identity thinking, which he developed most fully in *Negative Dialectics* (1966). A generation younger than Lukács, we elucidate the problem of history and freedom during Adorno's moment, which coincides with the expansion of the Great Acceleration, during a later stage of capitalist development (the decades immediately following WWII). Against this background, we contextualize and extend Adorno's critique of identity thinking in order to outline an immanent critique of contemporary environmentalism.

Chapter 3 provides a close reading of Moishe Postone's critical Marxian theory, which he advances through a critique of so-called traditional Marxism. A generation younger than Adorno, Postone's critique of traditional Marxism is an attempt to work through the implications of Marx's critical theory of modern society under different historical circumstances—namely, the reversal of state-centrism after 1973 without the attendant growth of international Marxian mass politics—circumstances which coincide with the present moment of the Great Acceleration. We extend Postone's critical Marxian theory to specify the socio-ecological domination underlying what is conventionally referred to as "economic growth." In doing so, we elaborate conceptual tools that are uniquely well-suited to more fully comprehend the links between economic progress and ecological deterioration. The chapter concludes by situating human–ecological transformation and contemporary ecological subjectivity in relation to the Anthropocene and helplessness.

Notes

1 Paul Crutzen advanced the idea of the Anthropocene in the prestigious journal *Nature* (Crutzen, 2002). The term was rapidly and widely adopted by scientists and by 2008 the Stratigraphy Commission of the Geological Society of London recommended that the Anthropocene be considered as the latest geological epoch in the Quaternary Period (Zalasiewicz and Williams, 2008). Evidence included anthropogenic-mediated patterns of: (1) physical erosion and sedimentation, (2) biological extinction, biomass change, and widespread changes in species range, and (3) changing surface earth chemistry, particularly in terms of ocean acidification due to increased atmospheric carbon concentrations. Based on these findings, the International Commission on Stratigraphy recommended an Anthropocene Working Group be established within the Subcommission on Quaternary Stratigraphy to determine the merits of incorporating the Anthropocene into the geological timescale (Zalasiewicz et al., 2014). The Anthropocene Working Group is set to report on its findings in 2016. The final decision will reside with the International Commission on Stratigraphy and its parent organization the International Union of Geological Sciences. A decision will likely not be rendered for some years after the Working Group's report is delivered.

2   While Crutzen (2002) and Steffen et al. (2011) note that one can find traces of Anthropocene-like periodization as early as 1873 it has only recently become the subject of debate at international geological meetings and in the popular media (Brown et al., 2013).
3   While the growth of environmental subjectivity during the Great Acceleration, for example, is acknowledged by Crutzen and his colleagues, the reasons they provide for its ineffectiveness is inconsistent and unconvincing. In one formulation they assert that "signs abound to suggest that the intellectual, cultural, political and legal context that permitted the Great Acceleration after 1945 has shifted in ways that could curtail it" (Steffen et al., 2007: 618) and that "humanity is, in one way or another, becoming a self-conscious, active agent in the operation of its own life support system" (Ibid.: 619). They suggest that this will precipitate a public debate over various approaches for dealing with global environmental degradation. How such a consciousness emerged in the first place and why it might be expected to spread is not explained. The possibility that such consciousness might in some way be ineffective in the face of previous attempts or even constitutive of it is similarly never considered. Elsewhere Steffen (2008) borrows an argument advanced by Fischer et al. (2007) claiming that such a debate will remain impossible until we confront "foundational issues" through a process of cutting deeply across the traditional barriers between the sciences and humanities. Here the environmental consciousness arising in the 1970s and 1980s appears to be lacking and in need of a new synthesis. The source of this new consciousness, however, seems to be located firmly within the very same disciplines that arose in and through the Anthropocene, albeit in a different collaborative network. Again, how such disciplinary knowledge is deemed to be free of the problems that gave rise to the Anthropocene in the first place is unclear. Finally, in Steffen et al. (2011: 862) the persisting inaction to climate change is attributed to the hegemony of neoclassical economic thought. Change, they contend, will neither be inevitable nor arise from new forms of interdisciplinarity, but rather is contingent on humanity disabusing itself of "the notion of human 'progress' or the place of humanity in the natural world is directly challenged" (Steffen et al., 2011: 862). Presumably the agency that would provoke such a "challenge" would come from the process of environmental degradation itself. Yet such an agency, which looks to an external catastrophe to generate subjectivity, bears an uncanny resemblance to the process that retrospectively characterizes the Anthropocene, namely one that proceeds through an unthinking and automatic process.
4   According to Steffen et al. (2011: 850), the genesis of the Great Acceleration "was clearly evident in the 1870–1914 period," although "the acceleration of these trends was shattered by World War I and the disruptions of the decades that followed." As Steffen et al. (2011) explain, these trends were only reignited in the post-WWII era.

DOI: 10.1057/9781137503886.0005

1
Georg Lukács (1885–1971) and the Critique of Reification: On the Dialectical Genesis of the Great Acceleration

Abstract: *This chapter situates Lukács' critique of reification (1923) in relation to the emergence of the Great Acceleration. We develop Lukács' critique through the issue of the increasing rationalization of industrial and administrative work in the early twentieth century. In doing so, we show how Lukács is able to relocate the continued relevance of Marx's insights with respect to the deeper structure of capitalist society in his consideration of the differential manner in which proletariat and bourgeois class consciousness approach the problem of social contradictions. We then discuss how, for Lukács, the overcoming of reification (or the failure to do so) has profound implications for how society comes to regard history and the possibility of freedom. The chapter concludes with a discussion of the significance of Lukács' critique for our understanding of the Great Acceleration.*

Stoner, Alexander M. and Andony Melathopoulos. *Freedom in the Anthropocene: Twentieth-Century Helplessness in the Face of Climate Change.* New York: Palgrave Macmillan, 2015. DOI: 10.1057/9781137503886.0006.

1 Marxism in crisis: social democracy on the eve of the Great Acceleration

Georg Lukács' (1923) *History and Class Consciousness* (*HCC* hereafter) comes on the heels of an important split within the political currents of Marxism at the beginning of the twentieth century. This split, which had been brewing within Europe's social democratic parties in the years leading up to WWI, marked the end of the project to unify the political interests of the working class under the aegis of the socialist parties of the Second International beginning in 1889 (Joll, 1968). This movement into mass political parties marked a distinctive transition not only in the form that working class ferment took but also in terms of the relation of these discontents with respect to the national state framework. If the activity of the working class at the beginning of the Industrial Revolution was notably antagonistic to the state—from its first identifiable amorphous form in the 1830s with the Chartist movement in England (Hobsbawm, 1962), to its central role in the pan-European revolutions of 1848 and through to the Paris Commune of 1871—its articulation within social democracy appeared to have a somewhat more ambivalent relationship.[1]

This transition into the period of the Second International bears significance on the question of the Anthropocene, particularly in considering the historical origins of the Great Acceleration from the crucible of the Industrial Revolution. While the emergence of liberal capitalism was remarkable to the extent that it appeared to occur at a distance and in opposition to the Absolutist states of the eighteenth century, with labor markets and the investment of capital being organized by no single authority—the so-called invisible hand in Adam Smith's moment—by the late nineteenth century both labor markets and capital were being integrated into a distinctive state form that Karl Marx characterized as "Bonapartist."[2] For Marx, the Bonapartist state was inextricably connected with the historical development of the Industrial Revolution since Smith's time and the manner in which the working class's political demands for a solution to the problem of unemployment triggered a social crisis: "The proletariat, the lowest stratum of our present society, cannot stir, cannot raise itself up, without the whole superincumbent strata of official society being sprung into the air" (Marx and Engels, 1978 [1848]: 482). Bonapartism presented itself in the wake of the pan-European uprisings of 1848, in which the general demand for democratic reform was drawn into sharp crisis when combined with the demands of the proletariat. The authoritarian Bonapartist state was a means of mediating the growing social instability generated by the emerging proletarian politics.[3] For Marx, the severity of Bonapartism had deeper implications, which the authoritarian state merely reflected, namely, the changed character of society as a whole. The Industrial Revolution, as evinced by Bonapartism, suggests something more profound than a mere technological shift as Anthropocene periodization presupposes (e.g., the use of coal for steam power): it marks a profound transformation within the very fabric of society.

As mentioned previously, the great mystery of the Anthropocene is how human beings, who emerged approximately 200,000 years ago, only become recognizable as a planetary force in the last 250 years. While the political expression of the proletariat may appear to represent only a narrow sectional interest within society, Marx grasps it as the key historical development through which we might recognize the deep, seemingly indiscernible social structure that took form in Industrial Revolution. For Marx, proletarian politics—if developed in the direction of consciously provoking and sustaining a crisis—would render social structure comprehensible and therefore enable individual actors to exercise agency in actively transforming history for the first time. The task of freedom in the nineteenth century—that of consciously recognizing and actively transforming social structure—was inextricably linked to the question of the political activity of the proletariat. Yet, the expression of these politics through Bonapartism revealed a second insight: the most symptomatic expression of the crisis of society—the proletariat—lacked the type of critical consciousness necessary to recognize how its activity was connected to social structure. Bonapartism, then, is an index of the inadequacy of proletarian consciousness: "it was the only form of government possible at a time when the bourgeoisie had already lost, and the working class had not yet acquired, the faculty of ruling" (Marx, 1993 [1871]: 53).

Marx sought to specify the *possibility* that consciousness of the deeper, crisis-laden structure of society could be developed from the most acute consciousness available to the proletariat at the time; the description of the objective laws structuring their lives outlined in the classical political economy of Adam Smith and David Ricardo (discussed in Chapter 3). To this end, Marx marshaled the insights from his critique of political economy to inform the subsequent political orientation of the proletarian movement toward the possibility that it might consciously direct a revolutionary crisis through which society could apprehend its deep contradictory structure.[4]

Yet, the proletarian politics of the Second International, which focused on the Social Democratic Party, fell well below Marx's horizon. Rather than increasing consciousness of the total character of society among the proletariat, these politics were far narrower, taking aim at the numerous, discrete features of social reality (Korsch, 1970 [1923]: 57). While such an approach drew significant reforms from the state,[5] it dulled any insight of the deeper and contradictory character of society and it certainly did not necessitate revolution. In fact, the exact opposite occurred as the Second International justified a narrowing political strategy that was readily integrated into the state. Consequently, by the beginning of the twentieth century the practical problem of the revolution languished within Marxist theory,[6] giving way instead to a tacit evolutionary view of politics in which workers would gradually assume control of the state by expanding the scope of reforms (i.e., within the existing social structure). While the deeper social crisis continued unabated—and on a much larger scale (e.g., WWI)—it no longer appeared as a logical product of a contradictory social totality, but as a "bolt from the blue"[7] (Marx, 2008 [1852]: 9); that is, as episodic, inexplicable, and irrational events that lay beyond the concerns of political program.

PHOTO 1.1   *Rosa Luxemburg and the Second International*

Source: Rosa Luxemburg during a speech at the International Socialist Congress in Stuttgart, August 1907. Significantly, the stage is not only accompanied by a portrait of Karl Marx (on the right), but the other founder of the German Socialist movement, Ferdinand Lassalle (on left). Photo: AKG-Images.

Lukács' *HCC* is an attempt to regain Marx's point of departure in and through the debates and subsequent political divisions that erupted among Marxists in the decade before WWI over the question of whether revolution remained the pivotal feature of Marxian theory and political practice (see Photo 1.1). The content of these debates is expansive, but for the purposes of contextualizing *HCC*, Lukács is particularly concerned with the question of whether the necessity of revolution had been superseded by the success in exacting reforms from the state (the debate over revisionism in Germany, ca. 1897–1902),[8] or whether trade union militancy, or "trade union consciousness," would simply reproduce capitalist forms if allowed to unfold on its own volition (the debates in Russia over economism, ca. 1894–1902).[9] In both of these debates, Lukács, following the Second International political figures Lenin and Luxemburg, recognized the consciousness of the worker's movement was constitutive not only of the conscious self-transformation of society—through the revolutionary overcoming of wage labor—but could also reconstitute wage labor in new social forms of domination that would become increasingly naturalized, or "reified." Reified consciousness, not only among the bourgeoisie but within the workers movement, meant any revolutionary politics would need to understand the "gradations within the class consciousness of workers" (1971 [1923]: 78) as it relates to the "ideological problems of capitalism and its downfall" (1971 [1923]: 84).

Lukács' critique of reification, which focuses on the intrinsic interrelations of the subject–object dimensions of the commodity form, highlights the mediation between what Marx referred to as the "inner logic" of the capitalist mode of production and the pattern of political and cultural life at a later stage of capitalist development (see Dahms, 2011: 3–44). Since this distinctive phase of capitalist development appears to anticipate the pattern of state capitalism that gave the Great Acceleration its form, Lukács' critique of reification is particularly relevant to the last century of the Anthropocene. Significantly, Lukács' critique does more than describe the course of history in a negative sense (i.e., to anticipate the subsequent pattern of state capitalism, or more generally, the helplessness that eventually comes to define the Anthropocene), it specifies the conditions under which society's relationship to history could be freed from its "rigid, reified structure" (1971 [1923]: 202). Perhaps Lukács' most profound point is to discern the possibility that social actors could make history rather than be dominated by it. In this sense, Lukács places reification at the center of the practical tasks for freedom in the Anthropocene.

This chapter develops Lukács' critique of reification through the issue of the increasing rationalization of industrial and administrative work in the early twentieth century. We show how Lukács is able to relocate the continued relevance of Marx's insights with respect to the central importance to any understanding of the deeper structure of capitalist society in his consideration of the differential manner in which proletariat and bourgeois class consciousness approach the problem of social contradictions. We end the section by discussing how, for Lukács, the overcoming of reification in and through increasingly class conscious proletarian praxis (or the failure to do so) has profound implications not only for how society comes to regard history, but also—more central to the issue of the helplessness that characterized the Anthropocene—the significance of such praxis for qualitatively changing history such that humans, for the first time, could develop freely. The chapter concludes with some brief remarks on the significance of Lukács' critique for our understanding of the Great Acceleration.

2 Critique of reification

The concept of reification derived from Lukács refers to a form of social life under modern capitalism in which human subjectivity is increasingly shaped in accordance with the objective commodity form. Lukács' critique of reification is not, however, simply a critique of detached, contemplative individual forms of bourgeois subjectivity. Rather, Lukács seeks to grasp reification as a process of social mediation between consciousness and social structure, particularly as he asserts: "reification is (...) the necessary, immediate reality of every person living in capitalist society" (1971 [1923]: 197).

Combining Marx's critique of commodity fetishism[10] in ways similar to, yet quite distinct from Max Weber's theory of rationalization, Lukács emphasized how reification fragments social life, and he analyzed reification in relation to processes of increased rationalization, which emerged alongside the consolidation of large-scale industry and investment banks at the beginning of the twentieth century. Lukács viewed the concomitant rise of bureaucracy within the economy, the so-called managerial revolution, which gave rise to increasingly complex corporate and state hierarchies of organization and control, as a new form of domination in capitalist society. The managerial revolution had direct ecological implications as well, as it marked an increasing technocratic dimension to the management of ecosystems, not only in terms of preventing soil erosion on agricultural land (as discussed in the Introduction), but also in terms of sustainable harvesting of forests and fisheries (Olver et al., 1995; Straka, 2009).

Lukács redirects Weber's analysis of rationalization, identifying it as a force working in concert with the abstraction embodied in the commodity form. Keeping with his categorial appropriation of Marx's mature theory, Lukács is concerned not only with the objective aspects of rationalization but also with the mediation between objective processes of rationalization and its subjective effects as structured by the commodity form of social relations.[11] On this basis, Lukács identifies two key (subjective and objective) changes resulting from the rationalization of work: (1) the severance of the "organic, irrational and qualitatively determined unity of the product" (Lukács, 1971 [1923]: 88), which in turn effects an (2) equally abstract fragmentation of the subject whereby the worker's activity "becomes less and less active and more and more contemplative" (Lukács, 1971 [1923]: 89). In an attempt to discern capital's immanent crisis in his moment, Lukács (1971 [1923]: 92) analyzes the effects of economic rationalization in terms of the dissimulating role played by the commodity form: "This rational objectification conceals above all the immediate—qualitative and material—character of things as things. When use-values appear universally as commodities they acquire a new objectivity, a new substantiality which they did not possess in an age of episodic exchange and which destroys their original and authentic substantiality."

Lukács explains the reconfiguration of consciousness in connection to the effects of large-scale industry and related processes of increased rationalization from the perspective of the worker:

> [T]he period of time necessary for work to be accomplished (which forms the basis of rational calculation) is converted, as mechanization and rationalization are intensified, from a merely empirical average figure to an objectively calculable work-stint that confronts the worker as a fixed established reality. With the modern "psychological" analysis of the work-process (in Taylorism) this rational mechanization extends right into the worker's "soul": even his psychological attributes are separated from his total personality and placed in opposition to it so as to facilitate their integration into specialized rational systems and their reduction to statistically visible concepts. (Lukács, 1971 [1923]: 88)

In this passage it is clear that, for Lukács, economic rationalization is not solely economic. Although capitalism appears to advance human ends, the abstract universal dimension of the commodity form of social relations is analogous with an increasingly dominating force expressed through the defining features of modern social life, such as bureaucracy, instrumental rationality, and atomization. Unlike manufacturing capital, the production of wealth under large-scale industry becomes cut-off from direct human labor time expenditure. This deepening process of reification is most acute in the bureaucratic work that expanded in Lukács' historical moment, where "even his [the worker's] thoughts become reified" as well as the very "faculties that might enable him [the worker] to rebel against reification" (1971 [1923]: 172). Thus, Lukács recasts the apparent necessity between state planning, large-scale industry, and capital, as well as industrial labor and high levels of productivity, in light of the social mediations between the subject–object dimensions of the commodity form, thereby elucidating the historical specificity (and social character) of these forms.

Although it appears to be total, the process of rationalization, according to Lukács, is nonetheless inherently incoherent. This incoherence is inherent to industrial capital because isolated phenomena, such as the increased specialization of the division of labor and the associated fragmentation of social reality, are governed by a strict rationality. The rigidity of rationality both structures and is structured by the fact that these processes are severed from the social totality from which they were produced and continuously interact. More specifically, the commodity form of labor under industrial capital becomes alienated from itself as a "reified," immediate, and isolated "surface-level" appearance misapprehended as being directly indicative of the social totality itself. Lukács argues,

> It is evident that the whole structure of capitalist production rests on the interaction between a necessity subject to strict laws in all isolated phenomena and the relative irrationality of the total process (...) This irrationality, this—highly problematic—"systematisation" of the whole which diverges *qualitatively and in principle* from the laws regulating the parts, is more than just a postulate, a presupposition essential to the workings of a capitalist economy. (Lukcás, 1971 [1923]: 102–103 [original emphases])

For Lukács (1971 [1923]: 103), the increased specialization and fragmentation of labor "has the effect of making these partial functions autonomous [so that] they tend to develop through their own momentum and in accordance with their own special laws independently of the other partial functions of society." At the same time, the process of increased fragmentation leads to "the destruction of every image of the whole" (Lukács, 1971 [1923]: 103). Lukács uses bourgeois economic analyses of crises as an example of the fundamental misrecognition and ossification of this inversion of social reality. "[T]he structure of a crisis," he explains, is "no more than a heightening of the degree and intensity of the daily life of bourgeois society" (Lukács, 1971 [1923]: 101). "Crises" are experienced as such because "the bonds uniting [bourgeois society's] various elements and partial systems are a chance affair even at their most normal" (Lukács, 1971 [1923]: 101). In other words, what appears to be objectively given, or "natural" (i.e., nonsocial), is in fact social.

For Lukács, such contradictions exemplify the crisis of capitalism and as such, "are not concealed 'behind' consciousness but are present *in* consciousness itself (albeit unconsciously or repressed)" (1971 [1923]: 59 [original emphasis]). Although "the proletariat shares with the bourgeoisie the reification of every aspect of life" (1971 [1923]: 149), these contradictions appear differently in the consciousness of the bourgeoisie than they do in the consciousness of the proletariat. The activity of the bourgeoisie historically brings capitalism into being—first, in establishing formal equality in labor ("free" wage labor) and then in universalizing the commodity form[12]—but this activity is what ultimately binds the workers as a class; what makes "the fate of the worker" become "the fate of society as a whole" (1971 [1923]: 91). Yet, from the perspective of the bourgeois, this deeper social structure is inaccessible since society appears to them as a series of objective economic laws; society is something they approach as individual capitalists, hence "the social principle and the social function implicit in capital can only prevail unbeknown to them and, as it were, behind their backs" (1971 [1923]: 63). As capitalists seek to expand their domain, capital too threatens to generate the independent political organization of the proletariat, bringing to the surface "all the hidden forces that lie concealed behind the façade of economic life" (1971 [1923]: 65). The attending social crisis renders bourgeois consciousness increasingly contradictory. The most paradigmatic example of such a crisis is the 1848 French Revolution (previously discussed) where bourgeoisie "freedom" was transformed into its opposite by adding a new element—the activity of the proletariat:

> Politically, it became evident when at the moment of victory, the "freedom" in whose name the bourgeoisie had joined battle with feudalism, was transformed into a new repressiveness. Sociologically, the bourgeoisie

did everything in its power to eradicate the fact of class conflict from the consciousness of society, even though class conflict had only emerged in its purity and became established as an historical fact with the advent of capitalism. Ideologically, we see the same contradiction in the fact that the bourgeoisie endowed the individual with an unprecedented importance, but at the same time that same individuality was annihilated by the economic conditions to which it was subjected, by the reification created by commodity production. (Lukács, 1971 [1923]: 61–62)

Hence, in spite of the reifying character of bourgeoisie consciousness, the contradictory social totality becomes discernible *to society* by the political activity of the proletariat. Reification is, thus, not an inevitable feature of capitalism, but one that can be overcome. As Lukács explains, the fragmented immediacy of reified thought can then be revealed as being "the consequence of a multiplicity of mediations" enabling "the fetishistic forms of the commodity system...to dissolve", such that "in the commodity the worker recognizes himself and his relations with capital" (1971 [1923]: 168).[13] But since the "fate of the proletariat" becomes "the fate of society as a whole" under capitalism, "the self-understanding of the proletariat is therefore simultaneously the objective understanding of the nature of society" (1971 [1923]: 149), which is inherently contradictory. Hence, proletarian politics become the vehicle through which society can achieve mastery of itself.

If proletarian consciousness has the *potential* to move through contradictions in the direction of social totality, bourgeois consciousness is *restricted* to a fragmented view and is unable to recognize contradictions as such. Consequently, rather than recognize crises as a product of society, bourgeois consciousness, as exemplified by the natural sciences, tends to isolate social phenomenon, considering them discrete and unconnected:

> The more highly developed it [modern science] becomes and the more scientific, the more it will become a formally closed system of partial laws. It will then find that the world lying beyond its confines, and in particular the material base which is its task to understand, *its own concrete underlying reality* lies, methodologically and in principle, *beyond its grasp*. (Lukács, 1971 [1923]: 104 [original emphases])[14]

Lukács here criticizes the Second International Marxist economist Tugan-Baranovsky[15] and his attempt to explain production in purely quantitative terms.[16] The formalism of bourgeois thought, according to Lukács, has political implications:

> The reified world appears henceforth quite definitively—and in philosophy, under the spotlight of "criticism" it is potentiated still further—as the only possible world, the only conceptually accessible, comprehensible world vouchsafed for us humans (...) By confining itself to the study of the "possible conditions" of the validity of the forms in which its underlying existence is manifested, modern bourgeois thought bars its own way to a clear view of the problems bearing on the birth and death of these forms, and on their real essence and substratum. (Lukács, 1971 [1923]: 110)

Reification is embodied in the political, economic, and cultural institutions of capitalist society and it is enacted by the individuals who inhabit this society. Indeed, the "birth and death of these forms" underlies a very profound historical context. According to Lukács, the failure to situate the fundamental problems of society within the historical context within which they emerged is itself explained with reference to the intrinsic relationship between social structure and subjectivity. Lukács explains that bourgeois thought exhibits a "double tendency," which is also characteristic of bourgeois society, and that it expresses this in an opposition between an objective material world and subjective consciousness:

> On the one hand, it [bourgeois thought] acquires increasing control over the details of its social existence, subjecting them to its needs. On the other hand it loses—likewise progressively—the possibility of gaining intellectual control of society as a whole and with that it loses its own qualification for leadership. (Lukács, 1971 [1923]: 121)

Lukács (1971 [1923]: 122) believes this problem is ultimately rooted in the division between theory and practice. Lukács' theory of praxis seeks to move beyond traditional subject–object epistemology. He indicates that both subject and object develop simultaneously through practice—and that this process is thoroughly dialectical. In other words, through praxis the subject both constitutes and is constituted by social structure. This practical activity, according to Lukács, is historically determinate.

It is on this basis that Lukács is able to ground his explanation of the antinomies of bourgeois thought, particularly the opposition between objective matter and subjective consciousness, in the relationship between social structure and subjectivity, a relationship reflective of the contradictory nature of modern capitalist society:

> [M]an in capitalist society confronts a reality "made" by himself (as a class) which appears to him to be a natural phenomenon alien to himself; he is wholly at the mercy of its "laws", his activity is confined to the exploitation of the inexorable fulfillment of certain individual laws for his own (egoistic) interests. But even while "acting" he remains, in the nature of the case, the object and not the subject of events. The field of his activity thus becomes wholly internalized: it consists on the one hand of the awareness of the laws which he uses and, on the other, of his awareness of his inner reactions to the course taken by events. (Lukács, 1971 [1923]: 135)

In capitalist society, "'nature' becomes highly ambiguous" (Lukács, 1971 [1923]: 136) since, with the progression of modern capitalist society, "nature" becomes increasingly socialized while "society" becomes increasingly naturalized. It is here that the significance of praxis in Lukács' critique of reification is perhaps most obvious. Regarding the contradiction between subject and object, Lukács explains,

> [T]he contradiction does not lie in the inability of the philosophers to give a definitive analysis of the available facts. It is rather the intellectual expression of the objective situation itself which it is their task to comprehend. That is to say, the contradiction that appears here between subjectivity and objectivity (...) is nothing but the logical and systematic formulation of the modern state of society. For, on the one hand, men are constantly smashing, replacing and leaving behind them the "natural," irrational and actually existing bonds, while, on the other hand, they erect around themselves in the reality they have created and "made," a kind of second nature which evolves with exactly the same inexorable necessity as was the case earlier on with irrational forces of nature (more exactly: the social relations which appear in this form). (Lukács, 1971 [1923]: 128)

Lukács claims Marx's method allows us to grasp the mediation of the historical totality in and through the immediate, fragmented aspects of reified reality. As he explains,

> [T]he essence of history lies precisely in the changes undergone by those *structural forms* which are the focal points of man's interaction with environment at any given moment and which determine the objective nature of both his inner and his outer life. But this only becomes objectively possible (and hence can only be adequately comprehended) when the individuality, the uniqueness of an epoch or an historical figure, etc., is grounded in the character of these structural forms, when it is discovered and exhibited in them and through them. (Lukács, 1971 [1923]: 153 [original emphases])

As we discussed earlier, critical recognition of reification implies the possibility of undoing reification through revolutionary proletarian praxis. Following Marx, Lukács contends that such a revolution would need to be led by the proletariat. But the extent to which bourgeois ideology predominated these politics it would serve to undermine the revolution and deepen reification. Yet, the question of the ideological maturity of the proletariat loomed large in 1923 after recent and decisive defeat of communist uprisings in Germany, Italy, and Hungary (of which Lukács was a key political figure). It is important to note, for example, that many of the figures in *HCC* who are exemplars of bourgeois consciousness (e.g., Tugan-Baranovsky, Strauve, Mach, Sombart, Bernstein) were, in fact associated with the primarily Marxist Second International, that is, the organized proletarian politics before WWI. As Lukács warns:

> The danger to which the proletariat has been exposed since its appearance on the historical stage was that it might remain imprisoned in its immediacy together with the bourgeoisie. With the growth of social democracy this threat acquired a real political organisation which artificially cancels out the mediations so laboriously won and forces the proletariat back into its immediate existence where it is merely a component of capitalist society and not *at the same time* the motor that drives it to its doom and destruction. (Lukács, 1971 [1923]: 196 [original emphases])

Yet, against the reified consciousness of these figures, Lukács is explicitly drawing on the example of the Second International "Orthodox Marxist" radicals, such as Rosa Luxemburg and Lenin, and their attempt to address long-since languished questions of revolutionary organization and theory. However, this revolutionary tide has passed. Its critical insights, as captured in *HCC*, failed to register into a renewed praxis as the working class became "imprisoned in its immediacy" and integrated into the new command economies that stoked the fires of the Great Acceleration. As alienation became second nature, the obstacles preventing such critical consciousness from being realized in practice became insurmountable (Lukács, 1971 [1923]: 70). "The most striking division in proletarian class consciousness," according to Lukács (1971 [1923]: 70–71), "is the separation of the economic struggle from the political one." Lukács (1971 [1923]: 71) argues that this division is rooted in "the dialectical separation of immediate objectives and ultimate goal and, hence, in the proletarian revolution itself." The dialectical relationship between immediate objectives and ultimate goal is tricky because, as Lukács (1971 [1923]: 71 [original emphases]) explains, the latter necessitates that the proletariat transcend itself: "the revolutionary victory of the proletariat does not imply, as with former classes, *the immediate realization of the socially given existence of the class*, but, as the young Marx clearly saw and defined, *its self annihilation.*"[17]

Yet, it is precisely this feature of proletarian politics and its capacity to overcome the most general feature of capitalist society—the proletariat itself—that causes Lukács to attach epochal significance to it. The question he takes up from Marx at the beginning of his essay, "Class Consciousness," in *HCC*—"*what is the proletariat* and what course of action will it be forced historically to take in conformity with its own nature" (1971 [1923]: 46 [original emphases])—has implications for the history of humanity, and by extension (in the Anthropocene), the planet.

> The "realm of freedom," the end of the "pre-history of mankind" means precisely that the power of the objectified, reified relations between men begins to revert to *man*...In other words, when the final economic crisis of capitalism develops, *the fate of the revolution (and with it the fate of mankind) will depend on the ideological maturity of the proletariat, i.e. on its class consciousness.* (Lukács, 1971 [1923]: 69–70 [original emphases])

The implication is that the scope of the historical transformation, which attends the confrontation with the contradictory totality of bourgeois society, would mark a transition without parallel in the Anthropocene. In fact, it anticipates a transition that would draw to an end "pre-history"; all history extending back to the beginning of the Holocene. This stunning conclusion comes at the very end of Lukács' painstakingly critique of reification. In bourgeois society, "historical thought perceives the correspondence of thought and existence in their—immediate, but no more than immediate—rigid, reified structure" (1971 [1923]: 202). In this way, reification conditions history because the *potential* trapped *within* history can never be regarded *as potential*, but rather as fixed. What reification means is that historical reality can only appear as empirical facts assembled rigidly in a structure that is deemed invariable (e.g., Tugan-Baranovsky). But from the perspective of totality "reality is by no means identical with empirical existence"; rather, "reality is not, it becomes" (1971 [1923]: 203). We contend that the question of becoming, which is synonymous with the question of freedom, has itself become even more inaccessible in the almost one hundred years since *HCC* was published. The deepening of reification has considerable implications for how we understand the historical reality that is represented in the Anthropocene.

3 Reification and the dialectical genesis of the Great Acceleration

Grappling with the problem of history and freedom in his moment, we approach Lukács' critique of reification as an attempt to name a key mediating process between the objective dimension (i.e., concrete human–ecological transformation) and the subjective dimension (i.e., the social conception of "nature") of the environment–society problematic. Here we emphasize Lukács' understanding of mediation as a theory of praxis. In this sense, "reification" attempts to name a key mediating process, constituted by the capitalist mode of production, through which people create structures that in turn dominate them (cf. Postone, 1993: 30–33). At the same time, reification *rewrites social reality* so as to inhibit these very same humans from consciously recognizing that this is indeed the case (Stoner, 2014: 632).

Although not concerned with the Anthropocene per se, Lukács grappled with the meaning of history and freedom in and through the emergence and development of the commodity form of social relations. His critique of reification is an attempt to specify the question of the possible supersession of the capitalist mode of production. In doing so, Lukács sought to discern the prospects for regaining revolutionary practice from existing reified consciousness within the proletarian movement; that is, by understanding the historical significance of the emergence and development of the commodity form of social relations and how this transition necessitates overcoming the commodity form as the primary task of freedom.

Lukács' focus on the subjective dimension of the commodity form elucidates the commodity not only as a form of constraint, but also, importantly, as a form of agency. This understanding of freedom during the genesis of the Great Acceleration contrasts sharply with that of Crutzen and his colleagues. Cutoff from its historical and dialectical genesis, the concept of the Anthropocene advanced by Crutzen et al. is unable to effectively grasp how a society that emerges from the Industrial Revolution can be both conscious of the degradation of planetary systems *and* seemingly powerless to do anything about it. Conceiving of the Anthropocene as an objective historical reality necessitates a subjectivity made possible only at the close of the twentieth century, one which uncritically reflects the apparent incapacity of society to self-consciously transform itself toward less ecologically destructive ends.

Unable to account for its own integration and active involvement in this development, the objective possibility of the Anthropocene indicates the degree of distance between a less ecologically destructive form of social organization and "the adequate understanding of the total situation" (Lukács, 1971 [1923]: 79). Following Lukács, the Anthropocene cannot be adequately understood in terms of its effects; rather, the Anthropocene can only be understood in terms of its objective possibility. Although significant, the proposed Anthropocene periodization misapprehends the genesis of the Great Acceleration, mistaking reified forms of appearance as directly indicative of an underlying historical dynamic. As a consequence, Crutzen and his colleagues effectively naturalize, or "reify" a historically specific form of appearance which coincides with the beginning of the Anthropocene.

Following Marx, Lukács grasps the social totality through the continuation of the social crisis generated by proletarian politics—a dynamic, which not coincidentally, is synchronous with the proposed Holocene/Anthropocene transition. Moreover, the problem of self-conscious social transformation in the context of capitalism took on a practical dimension for Lukács, which is something that becomes less and less clear with the failure of Marxist politics and the unfolding of the twentieth century. Many of the same forms survive this moment—social democracy, labor unions, communism—but their capacity to work through the subjective dimension of the commodity form ("working class consciousness") does not—their practical connection to the problem of freedom become dimmer. According to Lukács (1971 [1923]: 76), "As the product of capitalism the proletariat must necessarily be subject to the modes of existence of its creator. This mode of existence is inhumanity and reification." Hence, the importance of self-criticism on behalf of the political Left, which we aim to advance in the chapters that follow.

Notes

1 The German Social Democratic Party (SPD) of the Second International is an important case in point (Nettl, 1965). In principle, the SPD aimed to hasten revolution and the collapse of the existing form of society. But its steady growth after the repeal of the Anti-Socialist Law in 1890 did not lead to increasing revolutionary pressure, but rather a significant softening of its position. The softening first became notable following the Russian Revolution of 1905, after which the SPD appeared to let a corresponding

wave of labor unrest in Germany draw down. By 1912, after the SPD's electoral victory of 110 seats made it the largest party in the *Reichstag*, the party began taking steps to actively support the government. Finally, on August 4, 1914, as German troops entered Belgium, SPD deputies supported a motion to advance war credits to the Imperial German state that enabled it to effectively declare war. As Max Horkheimer (1978 [1940]: 101) remarks: "The murky relationship between Lassalle, the founder of the German socialist mass party, and Bismarck, the father of German state capitalism, was symbolic. Both aimed at state control. Government and opposition party bureaucrats from the left and right were pushed towards some form of authoritarian state."

2 As Marx would point out of Bonapartism: "The empire, with the *coup d'etat* for its birth certificate, universal suffrage for its sanction and the sword for its sceptre, professed to rest upon the peasantry, the large mass of producers not directly involved in the struggle of capital and labour. It professed to save the working class by breaking down parliamentarism, and, with it, the undisguised subservience of government to the propertied classes. It professed to save the propertied classes by upholding their economic supremacy over the working class; and, finally, it professed to unite all classes by reviving for all the chimera of national glory. In reality, it was the only form of government possible at a time when the bourgeoisie had already lost, and the working class had not yet acquired, the faculty of ruling" (Marx, 1993 [1871]: 53).

3 In France, a broadly constituted democratic uprising led to the formation of the Second Republic in February and the creation of universal suffrage but also a National Workshop program designed to absorb the unemployed in government-run businesses paid out tax revenues. Following the elections in April, however, the Workshops were closed. The closure led to a working class uprising in June that was violently suppressed. The resulting political instability lasted two years and resulted in *coup d'état* by the Republic's President, Louis Napoleon (nephew of Napoleon I), and the eventual dissolution of the Republic in 1852 and the severe curtailment of political liberty throughout French society. Paradoxically, the victory of liberal democrats in February 1848 ends with an appeal to the most illiberal dimensions of society as the only means to reestablish social order. As Marx famously wrote of the June suppression: "Bourgeois fanatics for order are shot down on their balconies by mobs of drunken soldiers, their domestic sanctuaries profaned, their houses bombarded for amusement—in the name of property, of family, of religion and of order" (Marx and Engels, 1978 [1852]: 603).

4 A founder of the International Workingmen's Association (First International) in 1864, Marx wrote a scathing critique of one of the founding documents of the social democratic party in Germany at the end of his life

(*Critique of the Gotha Programme*, 1875) and many of the key figures of the Second International (1889–1916) regarded themselves as direct followers of Marx (Korsch 1970 [1922]: 129–156).

5 These reforms included the key elements of the modern welfare state—for example, its national framework of old age pension and unemployment insurance. Significantly, the first of such reforms were advanced in Imperial Germany in the 1880s in parallel with the Anti-Socialist Laws that were not only banned from the *Reichstag*, but severely restricted the rights of Socialists to assemble and outright banned the ability of workers to form trade unions. These reforms were intended by the German Chancellor Otto von Bismarck to undercut the influence of social democrats by responding to their criticisms of society (Joll, 1974: 44).

6 As Nettl (1965: 73) observes of the German Social Democratic Party (SPD): "in all the years from 1882 to 1914 there was only one article in *Neue Zeit*, the theoretical organ of Social Democracy, on the subject of post-revolutionary society, and this treated the problem merely in a historical context—as a discussion of past millenarian societies. Even the revolution itself was little discussed; the technique of it not at all. The all-important topic of war was treated as an abstract evil, simply to be denounced. Interest was focused largely on contemporary questions of the day and their importance in the context of present Socialist attitudes, while broader questions affecting the SPD's future tended to be ignored."

7 According to Marx (2008 [1852]: 9–10), neither the liberal author Victor Hugo nor the socialist Pierre-Joseph Proudhon (who each represented two significant responses to the unexpected collapse of the Second French Republic into the Bonapart Empire) are able to grasp the deeper issue at work and, as a result, are forced to interpret historical transformation as exogenous to society—that is, as a "bolt from the blue" (Hugo), or as wholly identical to society ("the result of an antecedent historical development" (Proudhon)). Along this vein, Lukács (1970 [1923]: 158) writes: "As a result of its incapacity to understand history, the contemplative attitude of the bourgeoisie became polarised into two extremes: on the one hand, there were the 'great individuals' viewed as the autocratic makers of history, on the other hand, there were the 'natural laws' of the historical environment. They both turn out to be equally impotent-whether they are separated or working together-when challenged to produce an interpretation of the present in all its radical novelty."

8 The struggle over revisionism within the German Social Democratic Party (SPD) was exemplified in the debates between the revisionist Marxist theoretician Eduard Bernstein and the orthodox Marxist Rosa Luxemburg that took place between 1889 and 1903. Although the debate was largely among Party theorists, it touched on deep practical concerns

as the supporters of revisionism were broadly constituted in the party and were also most in touch with currents in German society. As Nettl (1965: 68) describes: "Bernstein's main supporters were the practical men of the party, the Trade Union leaders, practicing members of various professions who happened also to be Socialists." Although the SPD Executive and its supporters came down firmly against revisionism, such that it appeared defeated by 1903, few, such as Luxemburg and Lenin, recognized the debate as pointing to a deeper and unresolved problem within the proletarian movement. Lukács' contemporary Karl Korsch (1970 [1923]: 58) would explain revisionism "as the attempt to express in the form of a coherent theory the reformist character acquired by the economic struggles of the working class parties, under the influence of altered conditions." In a manner that strongly foreshadows *HCC*, Luxemburg's (1970 [1899]: 62) critique of revisionism was that it constituted "the theoretical generalization made from the angle of the isolated capitalist" in which "each organic part of the whole economy is seen as an independent entity."

9   Vladimir Il'ich Lenin was quick to generalize the debates in Russia over the need to work beyond "trade union consciousness" to a broader problem within working class politics internationally that threatened to reproduce the "the political reaction that has long reigned" if socialists avoided working through such problems. This is apparent in the very first footnote of his critique of economism in Russia, titled *What is to be Done?*: "At the present time (as is now plain for all to see), the English Fabians, the French Ministerialists, the German Bernsteinians, and the Russian Critics—all belong to the same family, all extol each other, learn from each other, and together take up arms against 'dogmatic' Marxism. In this first really international battle with socialist opportunism, will international revolutionary social-democracy perhaps gather sufficient strength to put an end to the political reaction that has long reigned in Europe?" (Lenin, 1975 [1902]: 74–75).

10 In the short introduction to the *Reification and the Consciousness of the Proletariat* in *HCC* Lukács (1971 [1923]: 83) indicates that he takes Marx's critique of political economy for granted, particularly the insight that "commodities must not be considered in isolation or even regarded as the central problem in economics, but as the central structural problem of capitalist society in all its aspects."

11 According to Postone (quoted in Blumberg and Nogales C., 2008), "Lukács takes the commodity form and he shows that it is not simply an economic category but that it is the category that can best explain phenomena like those that Weber tried to grapple with through his notion of rationalization, i.e., the increasing bureaucratization and rationalization of all spheres of life.

Lukács takes that notion and provides a historical explanation of the nature of that process by grounding it in the commodity."

12 "The first system of production able to achieve a total economic penetration of society" (Lukács, 1971 [1923]: 62).

13 The process of overcoming reification for Lukács, however, is not an immediate process. As it was for Marx, as well as for Lenin and Luxemburg, it involves working through stages in which, as Lukács explains, "the plenitude of the totality does not need to be consciously integrated into the motives and objects of action" (1971 [1923]: 198). In fact, even the seizure of state power by the proletariat and the reorganization of the economy on socialist grounds would only constitute a step in the direction of being able to clearly discern the contradictory social totality of bourgeois society (1971 [1923]: 208). Rather the process is continuous and involves the: *"constant and constantly renewed efforts to disrupt the reified structure of existence by concretely relating to the concretely manifested contradictions of the total development, by becoming conscious of the immanent meanings of these contradictions for the total development"* (1971 [1923]: 197 [original emphases]).

14 It is not our intention here to deal with the complex issue of reification and the natural sciences. In this particular passage we take Lukács to be providing an analogy from "outside, i.e., a vantage point other than that of reified consciousness" to the question of social totality (i.e., the natural sciences) (1971 [1923]: 104). On the issue of reification and the natural sciences, see Vogel (1996) and Feenberg (1999).

15 Tugan-Baranovsky's student, Nikolai Kondratiev, would later become well known for his theory of long-term cycles of economic expansion and contraction.

16 It is notable that Karl Korsch (1970 [1923]: 56–57) also identifies a similar pattern in the work of another prominent Second International theorist Rudolf Hilferding. Also see Moishe Postone's (2009: 97–101) critique of Giovanni Arrighi.

17 In a different formulation of this same point, Lukács (1971 [1923]: 80 [original emphases]) states, "*The proletariat only perfects itself by annihilating and transcending itself, by creating the classless society through the successful conclusion of its own class struggle.* The struggle for this society (...) is not just a battle wages against the enemy, the bourgeoisie. It is equally the struggle of the proletariat *against itself*: against the devastating and degrading effects of the capitalist system upon its class consciousness."

2
Theodor W. Adorno (1903–1969) and the Critique of Identity Thinking: The Great Acceleration as Historical Sedimentation

Abstract: *This chapter situates Adorno's critique of identity thinking (1966) in relation to the development of the Great Acceleration. We first establish the theoretical context of Adorno's critique of identity thinking with specific focus on his philosophy of history, concept of mediation, and the ways in which the shift toward a specifically negative dialectics is tied to Adorno's attempt to discern the problem of freedom in his time. Against this theoretical background, the latter half of the chapter illustrates Adorno's critique as a critical reflection on the social context of the 1960s in order to advance a critical, self-reflexive understanding of contemporary environmentalism.*

Stoner, Alexander M. and Andony Melathopoulos. *Freedom in the Anthropocene: Twentieth-Century Helplessness in the Face of Climate Change.* New York: Palgrave Macmillan, 20151 DOI: 10.1057/9781137503886.0007.

Lukács' critique of reification had a significant impact on the first generation of critical theorists associated with the so-called Frankfurt School in Frankfurt, Germany, not least among them included Theodor W. Adorno. Both Lukács and Adorno emphasized social contradictions within a historically specific totality. But unlike Lukács, whose theory of praxis sought to integrate class consciousness and revolutionary social change, Adorno exerted himself in ceaseless efforts to illuminate the *disintegration* of theory and practice, which he believed characterized his historical moment "after Auschwitz." So while Lukács' critique was of reification within the consciousness of workers, which necessitated working through these forms of appearance (i.e., the political debates of the Second International within Marxism over whether revolution was really necessary), with Adorno, this becomes a different problem, namely, how to maintain the impetus of critique as revolutionary practice ebbs. This shift in focus was no doubt the result of changed historical circumstances, as Adorno's critical theory is rooted in the historical experience devoid of a historical subject; hence, his abiding concern with the problem of history and freedom.

Adorno's attempt to grasp the meaning of freedom in his historical moment is perhaps most fully articulated in *Negative Dialectics* (1973 [1966]). Since any attempt to provide a general summary or definition of Adorno's negative dialectics is neither possible (Adorno tellingly referred to his approach as an "anti-system") nor desirable, this chapter focuses on a central aspect of Adorno's approach—namely, his critique of identity thinking, which registers and critically reflects on the history in which it is immersed. This history is marked by the crisis of bourgeois society and the political crisis of the proletariat, which we contend (gleaning insight from Adorno as well as Lukács) is not only the start of the Anthropocene but also the possibility of its complete redefinition. Whereas Lukács' critique of reification allows us to specify a form of social mediation via commodity production as a particular form of social practice that coincides with the emergence of the Great Acceleration, Adorno's critique of identity thinking exposes "administered" society and its apparent stabilization following WWII as the illusory means through which humans become masters of the Earth (in much the same that some can be vis-à-vis the products of their own alienated societies). The shift from Lukács' critique of reification (the failure of Marxism in the Second International) to Adorno's critique of identity thinking (state capitalism and the emergence of consumerism in the culture industry) must therefore be traced back into social structure so as to illuminate the meaning of the historical transformation this shift registers.

This chapter is divided into two parts. Part one, to which we now turn, establishes the theoretical context of Adorno's critique of identity thinking. Part two demonstrates Adorno's critique in action by situating it in sociohistorical context. We will begin by outlining Adorno's break with the Hegelian dialectic and the ways in which the shift toward a specifically *negative* dialectics is tied to Adorno's unwavering attempt to discern the problem of freedom in his time. We then discuss Adorno's critique of identity thinking as well as his philosophy of history and concept of mediation.

1 Dialectics as critique

In mid-November 1965, during a lecture on the possibility of negative dialectics, Adorno told his students that taking the dialectic in Hegel seriously requires moving beyond Hegel (Adorno, 2008: 28).[1] Adorno's own break with Hegel is grounded in what he considered truly dialectic in Hegel's philosophy[2]—namely, the core of a dynamic theory of experience.[3] Experience, according to Hegel, is an *active dynamic*—a process of recognizing and moving beyond the insufficient conceptual commitments of unreflected consciousness (O'Connor, 2013: 61).

Hegel's treatment of concepts is in accordance with the German idealist tradition (i.e., concepts are both categories of thought and categories of reality). The dialectical conceptualization of experience, as Hegel explains later, is also very different from that of contemporary science, which presupposes the possibility of disembodied analysis from a detached reference point:

> The way in which this movement has been brought about is such that it cannot belong to the fixed point; yet, after this point has been presupposed, the nature of the movement cannot really be other than what it is, it can only be external. Hence, the mere anticipation that the Absolute is Subject is not only *not* the actuality of this Notion, but it even makes the actuality impossible; for the anticipation posits the subject as an inert point, whereas the actuality is self-movement. (Hegel, 1977: 13 [original emphasis])

Awareness of the insufficiency of unreflected concepts is, as O'Connor (2013: 62) elucidates, an experience of contradiction. By actively revising its presuppositions about what the object is, consciousness transcends/ transforms itself. For Hegel, reason is what compels consciousness to respond to, rather than to ignore, a state of contradiction (O'Connor, 2013: 62). In other words, the actively questioning consciousness is impelled by reason. The disposition of what Hegel calls *determinate negation* involves recognizing the limits as well as the conditions of possibility of our conceptualizations. Determinate negation, in contradistinction to the abstract negativity of mere skepticism, involves recognition of the concept "as the result of that from which it emerges" (Hegel, 1977: 51). As Hegel (1977: 51 [original emphasis]) notes, "when (...) the result is conceived as it is in truth, namely, as a *determinate* negation, a new form has thereby immediately arisen, and in the negation the transition is made through which the progress through the complete series of forms comes about of itself." Determinate negation, then, implies self-correction, which is linked to both the knowing subject and the object of analysis (O'Connor, 2013: 63). For Hegel, insofar as the subject's conceptual commitments are challenged and reworked, the object itself is transformed:

> ...in the alteration of the knowledge, the object itself alters for it too, for the knowledge that was present was essentially a knowledge of the object: as the knowledge changes, so too does the object, for it essentially belonged to this knowledge (...) *Inasmuch as the new true object issues from it,* this *dialectical* movement which consciousness exercises on itself and which affects its knowledge and its object, is precisely what is called *experience* [*Erfahrung*]. (Hegel, 1977: 54, 55 [original emphases])

The negativity of Adorno's negative dialectics is rooted in Hegel's concept of determinate negation.[4] It is also Adorno's commitment to determinate negation that provides the basis for his critique of Hegel. For Adorno, Hegel's system, which involves a series of progressive steps as a rational requirement to overcome incompleteness (i.e., the movement from partial to absolute knowledge), falls short of what is essentially dialectical in Hegel's philosophy.[5]

According to Adorno (2008: 28), it is Hegel's "assertion that something can simultaneously be both a synthetic and an analytical proposition (...) that marks the point at which (...) we have to go beyond Hegel, if we are to take him seriously." Adorno breaks with the "positive" nature of dialectics in Hegel—in other words, with the notion that the whole, as the embodiment of all negations, is the positive. As Adorno (2008: 27

[original emphasis]) explains, "the fixed, positive point, just like negation, is an *aspect*—and not something that can be anticipated, placed at the beginning of everything." The fact that this premise is both what fuels the dynamism of the dialectic *and* what is supposed to emerge from it is precisely the point at which Adorno advances (negative) dialectics as critique.

Adorno's analysis and break with Hegel is other and more than a matter of logic, to be sure. The concepts Hegel is approaching are those of bourgeois society, though it is only with Marx that these concepts are illuminated as self-contradictory. The dialectic, therefore, approaches a totality (knowledge of bourgeois society as prehistory) that is not the one Hegel expected, which implies that this form of society must be overcome so that history might be possible.[6] As Thomas Huhn (2004: 5) explains, Adorno contends that the progressive arc of the dialectic and its integrative synthesis is not inevitable and that modern history is the measure of this failure. Adorno's push toward a specifically *negative* dialectic is conditioned by the fact that the dialectic historically has been suspended; hence, the opening statement of *Negative Dialectics* that philosophy lives on because the moment to realize it was missed (Adorno, 1973 [1966]: 3).[7] Significantly, the publication of Adorno's *Negative Dialectics* in 1966, in which he fully developed his critique of identity thinking, coincides with the explosion of the Great Acceleration following WWII.

According to Adorno (2008 [1965]: 7), the shift toward a specifically negative dialectics (as opposed to Hegel's systematic integrative dialectic alluded to earlier) is necessitated by the relationship between contradiction *in* the (dialectical) concept, on the one hand, and the social–historical content (whose essence *is* by virtue of its contradictions) to which the concepts refer, on the other (cf. Adorno, 1973 [1966]: 151–153). Indeed, Adorno's negative dialectics emphasizes this twofold meaning of the concept of contradiction, namely, the relationship between "a contradiction in the realm of ideas and concepts" and the fact that "the world itself is antagonistic in its objective form" (Adorno, 2008 [1965]: 9).[8]

1.1 Critique of identity thinking

According to Adorno, the relationship between concepts and the historical context to which they refer is reified under modern capitalist society. "[C]oncepts," Adorno (2008 [1965]: 23) explains, "are no longer measured against their contents, but instead are taken in isolation, so that people take up attitudes toward them without bothering to inquire further into the truth context of what they refer to." But such hypostasis comes at a price:

> This means that the less the mind possesses predetermined so-called substantial, unquestioned meanings, the more it tends to compensate for this by literally fetishizing concepts of its own devising which possess nothing that transcends consciousness. In short it makes absolutes of things it has created. And it achieves this by tearing them from their context and then ceasing to think of them further. (Adorno, 2008 [1965]: 24)

In other words, once the identity between concept and object has been reached, the need for critical self-reflection grows as the capacity for critique becomes impoverished. (It becomes impoverished because as reified consciousness the subject reflects upon concept and object as identical; in other words, the subject misapprehends concepts to be corresponding to the truth of what they identify).

This identitarian logic has its analogue in Marx's category of commodity, which renders both use-value and value identical, thereby concretizing two distinct, yet related, dimensions into a single form.[9] Exchange, which has come to be the measure of all things (Adorno, 1997 [1970]: 310), is, as Deborah Cook (2011: 19) explains, "the social model for what Adorno calls identity thinking." Attempting to grasp the relationship between concept and object within the reified matrix of modern society, Adorno's critique of identity thinking is an attempt to confront reification as a totalizing dynamic so as to discern the problem of freedom. For example, in directing focus toward what makes the concept of "society" (as something other and more than the sum total, or aggregate of individuals' interactions) possible, Adorno illuminates a historically specific form of self-generated domination:

> What really makes society a social entity, what constitutes it both conceptually and in reality, is the relationship of exchange, which binds together virtually all the people participating in this kind of society (...) Such a concept of society becomes, through its very nature, critical of society, in that the unfolding of the exchange process it refers to, objectively located within society itself, ends up by destroying society. To demonstrate this was really Marx's intention in Capital. Society, therefore, if it is to continue to reproduce the life of its members (...) must transcend the concept of exchange (...) by virtue of existing for others and being defined essentially as workers, human beings cease to be something existing in itself, a mere fact, but define themselves by what they do and by the relationship existing between them, namely, that of exchange. (Adorno, 2000 [1968]: 32–33)

Because an extensive exegesis adequate to the pertinent theoretical aspects of Adorno's work in this regard is well beyond the confines of the present chapter, we will draw on the work of Gillian Rose (1976, 1979) who provides an instructive gloss on Adorno's critique of identity thinking. Rose (1976) explains that concepts involve defining and, as such, are always contradictory. Concepts identify properties of an object, but properties do not exhaust what an object is. Identity thinking, which under capitalism is our normal mode of cognition, is in this sense totalizing[10]—it makes unequal things equal. There are three aspects of identity thinking: (1) pragmatic identity-thinking; (2) utopian identity-thinking; and (3) rational identity-thinking (Rose, 1976: 70).

Pragmatic identity-thinking refers to the "nature-controlling function of thought" (Rose, 1976: 70). The utopian aspect of identity-thinking refers to the fact that concepts also always relate to their objects by "the conditions of their ideal existence" (Rose, 1976: 70). Furthermore, under existing social conditions the utopian aspect of identity-thinking is antagonistic to itself. As Rose (1976: 71) explains, "For the concept to identify its object in this [utopian] sense the particular object would have to have all the properties of its ideal state." This condition marks the third aspect of identity thinking—rational identity-thinking. Although rational identity-thinking is constitutive of the utopian moment of identity thinking, it is problematic precisely because the utopian moment is barred from actuality due to the inherent violence of rational identity-thinking. In other words, under present social conditions it is impossible for the concept to identify its true object (Rose, 1976: 71). Rose suggests that Adorno confronts this problem by way of negative dialectics, which, as she explains, is a mode of cognition as non-identity thinking. Although non-identity thinking is different from identity thinking, it is neither external nor separate from identity thinking. For Adorno (1973 [1966]: 147), the possibility of a different mode of cognition is intrinsically tied to identity thinking: "Totality is to be opposed by convicting it of nonidentity with itself—of the nonidentity it denies, according to its own concept."

Adorno's negative dialectics, and his critique of identity thinking in particular, employs concepts to "break out" of the stranglehold of the concepts themselves. The "cognitive utopia," according to Adorno (1973 [1966]: 10), "would be to use concepts to unseal the nonconceptual with concepts, without making it their equal." That the whole is, in contrast to Hegel, quite literally "wrong" is what necessitates *negative* dialectics:

"In the unreconciled condition, nonidentity is experienced as negativity" (Adorno, 1973 [1966]: 31). (Negative) dialectics as critique, according to Adorno, is the "anti-method" most appropriate to a totalizing "administered" world. Not coincidentally, the aim of Adorno's theory is characterized by an uncompromising critical recognition of the present absence of social subjectivity: "Regarding the concrete utopian possibility, dialectics is the ontology of the wrong state of things. The right state of things would be free of it: neither a system nor a contradiction" (Adorno, 1973 [1966]: 11).

1.2 Negative philosophy of history

Adorno's 1964 lecture notes on "Progress or Regression?" bear the historical specificity of the atomic bomb and the resultant threat of mass planetary destruction. "Even if the murder of millions could be described as an exception and not the expression of a trend (the atom bomb)," writes Adorno (2006 [1964]: 4), "any appeal to the idea of progress would seem absurd given the scale of the catastrophe." For Adorno, following Hegel, history is meaningful insofar as it is the story of freedom. Yet, unlike Hegel, Adorno (2006 [1964]: 5) argued that the theory of historical progress, "as an advancement in freedom," was "highly vulnerable." Adorno (1998 [1962]) contends that in order to speak of progress the social conditions for progress (i.e., the actuality of freedom) would first have to be established. On this basis, Adorno (2006 [1964]: 10, 29–38) maintains that history is only objectively possible as philosophy of history and that, consequently, in order to do justice to the object of study we must be able to deal with the complex issues involved in mediating between the universal and the particular.

Adorno's philosophy of history pivots upon the *meaning* of the objective course of history vis-à-vis the particularity it ensnares. Adorno refers to this process as the "spell" of capitalist society.[11] As he explains,

> [W]hat Hegel calls the world spirit is the spirit that asserts itself despite people's wishes, over their heads, as it were. It is the primacy of the flow of events in which they are caught up, and it impinges on them no less than do the facts. Only it does so less painfully, and is therefore the more easily repressed. What is important here is that you should not regard this idea of the spirit prevailing over people's heads as a kind of speculative prejudice and hence dismiss it all too readily. It is important, I say, that you should realize that this is a process in which what prevails always passes not merely over people's heads, but *through* them. (Adorno, 2006 [1964]: 25 [original emphasis])

This has implications for comprehending history, as Adorno's (2006 [1964]: 29) critique of historicism makes clear: "Facts that have become a counterweight to mere illusion (...) reinforce the impression of mere illusion." The power of facts to penetrate illusion, via the disenchantment of the Enlightenment, has itself become a historical illusion. As the world is divested of meaning, facts are mistakenly taken to be something other—*something more*—than what they actually are. Through this process, which Adorno tellingly refers to as "socially necessary semblance," positivism emerges and gains a momentum of its own. As such, Adorno (2006 [1964]: 11) argues that both the defense and critique of the importance of historical "facts" presuppose that historical processes have some sort of meaning. According to Adorno, the meaning of history can only be discerned by critically reflecting on the unfreedom of the present so as to discern the possibility that history (i.e., freedom) may begin.

Adorno advances his philosophy of history via a critique of history. His "historical methodology" therefore operates at two different levels simultaneously. Adorno engages a critique of positivism while at the same time advancing a negative philosophy of history in and through this critique. According to Adorno, the primacy of the objective course of history takes precedence over the particular individuals who create history because the overall process of history operates in and through them as "isolated" particulars. It is not that the objective course of history *should* be primary; rather, the objective course of history *actually is* primary, and we must be able to take this into account if we are to do justice to the study of historical reality. Adorno contends that positivism, including the dominant historical methods that focus on isolated specificity, is false because it does not recognize the ways in which its mode of historical inquiry is itself determined by history and, as such, says nothing about history. In other words, our comprehension of historical facts is itself a product of the objective course of history, which penetrates our social being. Adorno's central point here is that history is impossible if we fail to recognize this.

How, then, can we say anything meaningful about history? Adorno is fully aware of this dilemma. Indeed, his emphasis on the non-identical is a response to and immanent critique of this problematic. Adorno follows Benjamin in engaging the primacy of the objective course of history, but rather than emphasizing a one-sided focus on the totality, Adorno aims to capture the *mediation* of the totality in the particular. As Adorno (1973 [1966]: 303) explains, "totality over phenomenality is to be grasped in phenomenality." The key, for Adorno, is to see how the particular *expresses* the totality. Adorno deals with the problem of the meaning of history and is able to root his critique in the possibility of emancipation while offering an explanation of history that is able to account for nonlinearity—all with his concept of *mediation*.

In his 1964 lecture on the concept of mediation, Adorno explained to his students what must, at the very least, be included in an adequate explanation of historical factors:

> In short you need to grasp the complexity of the pattern, by which I mean the overall process that asserts itself, the dependence of that global process on the specific situation, and then again the mediation of the specific situation by the overall process. Furthermore, in addition to understanding this conceptual pattern, you need to press forward to the concrete, historical analysis I have hinted at and that goes beyond the categories I have been discussing. (Adorno, 2006 [1964]: 37)

Adorno's analysis of historical content employs an interpretive sociology, an approach he likened to that of physiognomy. From this perspective, the sociologist must study surface phenomena as manifestations of social structure (see, e.g., Adorno, 1978 [1938]). As Adorno (1976 [1969]: 32 [original emphasis]) explains, "In sociology, *interpretation* acquires its force both from the fact that without reference to totality—to the real total system, untranslatable into any solid immediacy—nothing societal can be conceptualized, and from the fact that it can, however, only be recognized in the extent to which it is apprehended in the factual and the individual."

2 Human–ecological transformation and contemporary ecological subjectivity

Before we situate the foregoing theoretical discussion in relation to sociohistorical context, it will be helpful to restate the parameters of the environment–society problematic. As outlined in Chapter 1, the environment–society problematic refers to the process in which societally induced ecological degradation has compounded in proportion to our awareness of these problems. Here we deciphered an objective dimension (actual, concrete human–ecological transformation) and a subjective dimension (the social conception and understanding of the natural environment), which allowed us to further specify an elusive dynamic wherein the synchronicity of each dimension comprises a single process whose unfolding throughout the latter half of the twentieth century appears increasingly paradoxical. In the remainder of this chapter, we will employ Adorno's critique of identity thinking to illuminate this otherwise elusive paradoxical process. We will explain how the reconfiguration of society and power relations following WWII—a context Adorno referred to as "administered" society—has its correlate in the form of consciousness articulated by the so-called new social movements in the late 1960s, including the contemporary environmental movement. We will argue that it is the identity thinking in environmentalism that has been the movement's greatest stumbling block. With identity thinking, environmental discontents are readily integrated into the capitalist system and are thereby used merely as a means to perpetuate the existing state of affairs. As we shall see, the Anthropocene too is strongly conditioned by the "subjectivity" of the environmental movement that emerged alongside Adorno's negative dialectics during the 1960s.

2.1 "Administered" society

Addressing the German Sociological Conference in 1968, Adorno warned his colleagues that the debate over the proper definition of post-WWII society and power relations—"late capitalism" or "industrial society"—was nonsensical insofar as it entertained, rather than critically reflected on, the continuation of social domination. As he explained:

> Concepts such as 'exchange-society' have an objective reality and a universal coercive force that goes beyond the facts and that cannot always be translated into operationally defined realities. Both tendencies must be resisted. In this sense our subject, late capitalism or industrial society, testifies to our intention to practice self-criticism in the spirit of freedom. (Adorno, 2003 [1969]: 113)

Social domination which had previously (during the 1930s) become "an integral part of human beings" (Adorno, 2003 [1942]: 109) was transformed and reasserted in the late 1960s and early 1970s. The formation of the welfare state in the mid-twentieth century, for example, while celebrated by liberals, was, in Adorno's (2003 [1942]: 105) words, merely "the system's consciousness of the conditions that enable it to be perpetuated" (where the "system" being referred to is domination itself).

Adorno formulated the transformation of society and power relations that became increasingly visible, yet less and less recognizable, following the end of WWII in terms of a totalizing "administered world." Adorno elaborates the insidious, seemingly non-contradictory nature of this social configuration, which has its correlate in a regressive form of consciousness, in the following manner:

> The idea that the forces and relations of production are one and the same today, and that the notion of society can be easily constructed solely by reference to the forces of production, is the current shape of socially necessary illusion. It is socially necessary because elements of the social process that were formerly separate—and this includes living human beings—have been brought down to a common denominator. Material production, distribution, and consumption are administered jointly. Their boundaries flow into one another, even though earlier within the overall social process they were at once different from one another and related, and for that reason they respected what was qualitatively different. Everything is now one. The totality of the processes of social mediation, which amounts in reality to the principle of exchange, has produced a second, deceptive immediacy. This enables people to ignore the evidence of their own eyes and to forget difference and conflict or repress it from consciousness. But this consciousness of society is illusion, because while it does justice to the process of technological and organization standardization, it overlooks the fact that this standardization is not fully rational but remains subject to blind, irrational laws. No overall social subject exists. (Adorno, 2003 [1969]: 124)

In the following pages we will discuss some of the political–economic shifts indicative of "administered" society by drawing on examples from the American context and the post-WWII reconfiguration of business–labor–government relations that took hold during this time (see Dahms, 2000, 2006).

In the American context, this post-WWII regime emerged through organizational linkages that had been established during the 1930s (e.g., in the changing relations between large investment banks and the federal government).[12] These organizational linkages, in turn, facilitated the economic boom that is typically associated with the so-called golden age as well as the post-WWII spike in environmental degradation. Although the New Deal administration put an end to the U.S. tradition of federal minimalism, its policies were necessity in order to save American society from the destructive effects of finance capitalism. However, it was WWII that brought the U.S. economy out of economic depression, specifically, the heightened need for rapid and large-scale armament production, which pushed further the growth of large-scale economic organizations.

In fact, the golden age of capitalist development was less spectacular in America because, as Hobsbawm (1994) has convincingly argued, it was a continuation of the economic expansion of the war years (see also Webber and Rigby, 1996). Indeed, the war effort intensified what Kim McQuaid (1994) refers to as big business–government symbiosis. State development and war making reinforced one another, as each evolved through a complex web of institutions and scientific expansion (McLaughlan, 1992; Mills, 1956). Amid growing Cold War tensions, conservatives saw the potential to seize control of the situation and reign in labor. Meanwhile, another configuration of power was in the works—one involving the political–economic alignment of the Business Committee (or BC) and the Committee on Economic Development (or CED), two business organizations that were able to take significant control of post-WWII reconstruction (McQuaid, 1994: 18–47).

What had appeared to be a potentially destabilizing atmosphere in post-WWII American society turned into corporate endorsement of Keynesianism and conservative power realignment. Labor's support of the New Deal administration's social democratic affinities was gradually inverted from 1945 to 1980 as postwar prosperity led to a refurbishing of classical liberal ideology characteristic of American conservativism (Lipset, 1997). In America, where socialism never sustained a large number of adherents, politics became even more narrowly defined during this period. The American political Left was eclipsed by liberalism by the 1950s, and any qualitative distinction between the Left and the Right was effectively flattened out (Hodgson, 1976: 89). However, Cold War liberalism and its hegemony fluctuated during the 1960s–1970s. This transition is marked by widespread social discontent, as evidenced by the crisis of the Civil Rights movement (peaking in 1965), the crisis of the Vietnam War, and the emergence of a host of "new" social movements, including the contemporary environmental movement.

During the 1960s, the Left became increasingly involved in environmental issues. Like the "old" social movements of the past, the contemporary environmental movement is divided internally between so-called radical and mainstream adherents. Today, for example, environmental politics is still situated in an uneasy relation to the Marxian Left. On the one hand, the growth of the environmental movement in the 1980s, particularly in Europe, marked the sharp migration of people drawn to Marxism in the 1970s to Green politics (Berman, 2005; Bookchin, 1986 [1969]; Biehl, 2012; Kivisto, 1986; Mewes, 1983). On the other hand, a common theme of environmentalism is to impose limits to growth, sometimes expressed in conservative sentiments against technology, urbanization, and cosmopolitanism, things that the Marxian Left historically took to be signals of progress. The radical/mainstream division has characterized the contemporary environmental movement since its inception. In the United States, this division was visible in the development of the antinuclear movement, one of the most immediate precursors of the contemporary American environmental movement.

Contemporary environmentalism was born in the post-WWII period of "stability" amid a generational and ideological shift marked by the questioning of authority. The deployment of the atom bomb brought to light for the first time on a massive public scale the possibility of civilizational collapse. This new prospect was joined by writings that contended that the free enterprise system was based on a flawed understanding that humans were somehow separate from the biophysical world. Authors such as Osborn, Vogt, Bookchin, Commoner, and Marcuse criticized society's misrecognition of the environmentally destructive effects of industrial society (Opie, 1998: 405). The targets of environmental discontent articulated by contemporary environmentalism—namely, technology and affluence—expressed the fact that, after nearly two decades of unprecedented economic growth, the industrial mode of production was no longer adequate. This criticism is historically specific, as the post-WWII form of "economic growth"[13] had expanded to such an extent that it became possible to question its necessity (see, e.g., Gorz, 1967; Nicolaus, 1968). Today, environmentalism continues to express discontent with the growing inadequacy of technology and affluence (e.g., "limits" to growth), as well as the administrative state more generally in addressing global environmental problems.

Yet, the social crises of the 1960s,[14] including the crisis associated with the rise of contemporary environmentalism, were readily stabilized. The success of the post-WWII configuration of business–labor–government relations, whose organizational forms were laid in the 1930s, produced new environmental threats that could not be conceived of in purely economic terms. Following WWII, the scale and scope of technological development expedited unprecedented levels of resource extraction, productivity, and global capital flows. *Although the post-WWII spike in environmental degradation gave rise to new environmental needs and desires, the environmental discontent expressed by contemporary environmentalism failed to engender changes in social structure conducive to moving beyond the societally induced environmental degradation which characterizes this period.* In fact, the exact opposite has occurred, as the growth of environmentalism throughout the 1970s and 1980s coincided with the advent of neoliberal global capitalism whose penetration continues to define our moment, despite growing efforts to ameliorate humanity's predictable ecological collapse. In this sense, environmentalism's apparent revolt gave rise not to a social crisis, but to the perpetuation of capital.[15]

The failure of contemporary environmentalism registers the deepening of a process—of which the Second International was an instantiation—whereby surface expressions are taken as natural; and hence, the social discontents of the 1960s, including those expressed by environmentalism, are readily incorporated into this system. Indeed, it is through this process that the environment–society problematic remains paradoxical and elusive. Although the environmental effects of the post-WWII period are becoming increasingly visible, solutions to these problems continue to be put forth in terms of the problem itself—and apparently without the potential for being otherwise.

2.2 Toward a critique of contemporary environmentalism

Why did the rise of contemporary environmentalism, instead of attaining its objective, shift the problem to other registers?[16] The answer to this question revolves around the identity between society and its discontents. Although the widespread discontents in the 1960s, including those expressed by the environmental movement, register the need for social transformation, it is through the identity thinking in these movements that such discontents are integrated into the capitalist system. With identity thinking, environmental discontents are used as the means to perpetuate capitalism. Consequently, without a critical, self-reflexive understanding of their own historicity, the capacity of these discontents to clarify social structure (and its entwinement with ecological degradation) is rendered impotent.

The emergence of what we call *contemporary ecological subjectivity*—a specifically post-WWII social conception and understanding of the natural environment[17]—was contingent upon the reconfiguration of post-WWII society and power relations and the ecological degradation this administered form of society facilitated. During the post-WWII, Cold War period, state managers increasingly viewed science and technology as the means for continual preparedness for war. This corresponds to what McLaughlan (1992) describes as a shift from "geopolitics" to "technopolitics" in which control of technology becomes the primary means for exercising and expanding state power. Similarly, scholars have detailed the ways in which Cold War military–industrial complexes produced the knowledge that facilitated the unprecedented expansion and control of the planet during this time (Doel, 2003; Edwards, 2010, 2006; McNeill and Unger, 2010).[18] Indeed, as these researchers are keen to point out, the sophisticated nature of our contemporary understandings of the biophysical world is one of the most significant, yet often overlooked, effects of the Cold War.

The connections between the Cold War boost in scientific knowledge, ecology, and the emergence of contemporary environmentalism in the late 1960s and early 1970s are intricate. According to Paul Edwards (2010, 1996), without this knowledge production, it is unlikely that awareness of global climate change would have occurred as rapidly as it did. Indeed, the tremendous increase in such knowledge production played a key role in facilitating what we refer to as contemporary ecological subjectivity, as many Cold War technologies would later play a significant role in establishing how we know what we know about global environmental problems (Taylor and Buttel, 1992).[19]

PHOTO 2.1 *Cold War technology and environmental subjectivity*

Source: An artist's rendition of the next Landsat satellite, the Landsat Data Continuity Mission (LDCM) (Landsat 8), that was launched in Feb 2013. Although Landsat 1 was launched at the height of the Cold War in 1972, in part as a way to forecast potential wheat yields in the U.S.S.R. before harvest, it has developed into an essential tool for monitoring deforestation and rates of habitat change for wild species. Photo: NASA.

Although examples abound, here we briefly mention two technological developments which, taken together, illustrate the significant role of Cold War technology in the development of contemporary ecological subjectivity. The first is satellite photography—perhaps the quintessential Cold War invention—which brought the finitude of the spherical Earth into view from space for the first time in 1968. The second is systems analysis, which, as Amadae (2003: 41) explains, "originated as tools used by the aircraft trade to prepare contract bids for the U.S. Air Force" and soon became a mainstay at the RAND corporation where it was promoted and utilized as "a comprehensive and rigorous science of war" (Amadae, 2003: 43). The infamous "missile gap" between the Soviet Union and the United States during the beginning of the Cold War and America's strategy of containment, for example, were both based on prescriptions derived from systems analysis. Arguably, the idea of "limits" to growth and the related notion of sustainability—dominant elements of contemporary environmentalism—would not have been possible without satellite photography (shown in Photo 2.1) and systems analysis, both of which are products of the Cold War. This is not to say that these technologies are somehow "bad," in and of themselves, for there is no doubt that our knowledge about global climate change would be impossible without these technologies (though this does raise the question of the relationship between the centrality of science in identifying global environmental problems and sociohistorical context).

The more important issue is how a technological critique, which is often articulated by environmentalists, is plausible when avenues for society to work through reification are gone. Here it is important to bear in mind Adorno's point regarding the administered world. The context of the 1960s discontents is this "system"—not only technology or affluence, or even the administrative state more broadly, but *all* social phenomena "are so completely mediated that even the element of mediation is distorted by its totalizing nature. It is no longer possible to adopt a vantage point outside the hurly-burly that would enable us to give the horror a name; we are forced to adopt its inconsistencies as our starting point" (Adorno, 2003 [1969]: 124). Although contemporary ecological subjectivity is mediated and produced by this context, liberalism and Cold War technology are non-identical to contemporary ecological subjectivity. When liberalism and Cold War technology are the targets of environmental discontent they become objects of criticism, which are then readily integrated into the capitalist system. Similarly, environmentalism does not recognize its conditions of possibility and therefore cannot recognize itself as part and parcel of that which it attempts to overcome. Consequently, environmentalism remains identical with society while appearing opposed to it. It is on this basis that environmentalism remains bound to reproduce the existing state of affairs even as its discontents point beyond this context.

Yet, as Andrew Feenberg's (1996) re-reading of the debate between Paul Ehrlich and Barry Commoner in the 1970s indicates, contemporary ecological subjectivity can take various forms—from Ehrlich whose biological focus on overpopulation effectively depoliticizes environmental politics by delimiting its sphere to personal action, to Commoner who sought solutions to the environmental crisis not in individual action but through collective control over society's political and economic institutions—the actuality of environmentalism is but one of many possible forms. That contemporary ecological subjectivity ultimately takes a form in which discontents are identical with, and therefore readily integrated back into, the capitalist system—despite its apparent predetermination—is not inevitable. Feenberg's gloss on the history of contemporary environmental politics is significant because he draws attention to important political issues that continue to plague the contemporary environmental movement (e.g., class, race, and national differences). And yet, these issues remain only as their historical meaning appears ossified with the unfolding of the twentieth century. Indeed, the possibility that contemporary ecological subjectivity might take a different form—that is, a form non-identical to existing social conditions—is entirely foreign to most.

The forms of environmental politics that arose in the 1970s revolved around the idea that government must intervene as environmental regulator (Opie, 1998: 405–406). The first Earth Day, held in the United States in 1970, is an excellent example of these politics and the ways in which environmentalism was incorporated into the administrative state. One of the key organizers and proponents of the event, Wisconsin Senator Gaylord Nelson (shown in Photo 2.2), stressed the increasing importance of the environmental agenda in an attempt to distance it from so-called radical New Left countercultural activists (Gottlieb, 2005 [1994]: 148–149). Whereas New Left environmentalists viewed Earth Day as a betrayal, President Richard Nixon saw the event in terms of potential voters and shortly thereafter proclaimed the environment a key issue, glorifying the new environmentalism as a noble cause to be supported

(Gottlieb, 2005 [1994]: 152), which he later legitimated through the creation of the Environmental Protection Agency (EPA). As Nixon sought to link his presidency and administration to technology-based solutions to environmental problems (Gottlieb, 2005 [1994]: 152), the American business community also began to embrace environmentalism (likely due to the fear of possible environmental legislation (Gottlieb, 2005 [1994]: 153). Although a number of large-scale corporations gave financial contributions to help organize the first Earth Day, by the early 1970s most big businesses began to associate themselves with environmentally friendly pollution control technologies.

According to Blühdorn (2013: 18), although the criticism mounted by early environmentalists "was complemented by the vision of, and desire for, a comprehensively different society (and modernity)" this political vision did not survive the 1970s, as environmentalism became increasingly integrated—for example, as modernization, nationalism, occupational health and safety, and the EPA. Regulation has turned into endless conferences that get nothing done, and protest does not have clear goals (and if they did, it is even less clear what would be done with these goals). Although the ideology that justifies engagement in capitalism has clearly changed as a result of the integration of environmental discontent into the capitalist system (e.g., the immense expansion of "green consumerism" in recent years), such integration comes at a price. As Boltanski and Chiapello (2005: 29) explain with regard to the disarming of critique in the late 1960s and early 1970s, "The price paid by critique for being listened to, at least in part, is to see some of the values it had mobilized to oppose the form taken by the accumulation process being placed at the service of accumulation."

Approaching this problem requires understanding the political forms emerging from the 1960s, a period that gives rise not only to the environmental movement, which affirms the static character of nature, but also postmodernism, which resolutely rejects this nature and all other essentialist claims (Biro, 2005). In fact, these two seemingly opposite approaches have common origins in the 1960s New Left, which fragmented in the 1970s into a "panorama of single-issue movements" (Berman, 2005: 65) including environmentalism, anti-nuclear and peace politics, identity politics, and the championing of struggles for self-determination and emancipation in former colonial states. Although there were attempts to unify these causes within the sphere of parliamentary politics, for example, in Green Party politics, this current appears to have

PHOTO 2.2 *Earth day and the environmental movement*

Source: The front page of the newsletter of the Wisconsin Senator Gaylord Nelson in May 1970. Nelson was one of the key organizers of nation-wide teach-ins on environmental issues on April 22, 1970. The event was observed at thousands of universities, colleges, and primary and secondary schools across the United States. In Manhattan the largest Earth Day gathering saw one million people participate in a march along Fifth Avenue. Photo: Wisconsin Historical Society.

reached significant limits within only 25 years. In its founding country of Germany, the Green Party has begun to move away from founding positions around pacifism and opposition to the nuclear energy industry and has supported neoliberal reforms to the German economy, making their policies difficult to distinguish from mainstream parties (Berman, 2005; Scott, 1990).

Current social movements engaged with environmental issues face a similar "despair with regard to the real efficacy of political will, of political agency" (Postone, 2006: 108). Blühdorn and others express this in terms of the "post-ecologist turn":

> [P]ost ecologist politics holds that since the late 1980s the normative foundations of eco-political communication have comprehensively changed. It suggests that in contemporary capitalist consumer democracies both the ecologist critique of modernity and the ecologist belief in a comprehensively different society have become largely exhausted. This exhaustion of the ecologist belief system is, arguably, due to a comprehensive value- and culture-shift, the *post-ecologist turn*, triggered, *inter alia*, by: the gradual normalization of the environmental crisis; the depoliticisation of the ecologist critique and the techno-managerial reframing of environmental issues within the paradigm of *ecological modernization*; the diversification—and thus relative weakening—of eco-political values and imperatives; and the becoming prevalent of ideals of identity, self-determination and self-realisation. (Blühdorn, 2013: 19 [original emphases])

But as people have become even more aware of societally induced ecological degradation since the 1970s—that is, as degradation becomes more apparent its form of political mediation does not lead to any sense of deep structure, but to an even more fragmented sense of the world—we wind up with a form of politics that essentially does nothing, leading to a widespread sense of helplessness:

> As the resolve to sustain, at least for the time being, what is widely recognised as unsustainable is the central characteristic that distinguishes contemporary eco-politics from earlier phases, the theory of post-ecologist politics conceptualises eco-politics in advanced post-industrial societies as the politics of unsustainability. Whatever its declared commitments, this politics of unsustainability is no longer powered by the ecologist attempt to change individual lifestyles and societal structures in such a way that environmental integrity may be sustained and ecologist visions of authentic social well-being achieved. Instead, its primary concern is to manage the inevitable consequences, social and ecological, of the resolve to sustain the established order. (Blühdorn, 2013: 20)

As we return to elaborate in the following chapter, it is this condition of helplessness that defines the present moment of the Great Acceleration. The Anthropocene naturalizes capital in history, in part because it remains incapable of elucidating the connection between the failure of revolution and the subsequent rise of state-centric capitalism. As a result, the paradoxical process whereby environmental degradation is compounded in proportion to our awareness of these problems remains elusive. The connection between Cold War technology and contemporary ecological subjectivity discussed earlier is indicative of the ways in which socioecological domination not only takes priority over, but also works through (and is thus perpetuated by) our conceptions and understanding of the natural environment. Adorno's vexing account of the actuality of historical transformation (existing the world over through global capitalism in addition to other systematic means) as well as in the guise of its weird and distorted mythological ideal self-understanding (its ideological existence) is particularly apt in light of the historical failure of contemporary environmentalism. Given the growing concern for the well-being of the planet in recent decades, particularly in the face of climate change, it is important to bear in mind that contemporary ecological subjectivity is not actual (critical) recognition of worldwide ecological crises if it is confined to the form of appearance of growing environmental attention and concern (made possible by increasingly sophisticated measurement technologies) without recognizing itself as such. With identity thinking, environmental discontents are readily integrated into the very same system of production driving the Great Acceleration. Actual (critical) recognition of today's worldwide ecological crises would mean recognizing this thought itself as a form of reified consciousness, so that the inherently irreducible, emancipatory potential contained within might be unleashed.

Notes

1 Adorno's 1965–1966 lectures, subsequently published in English by Polity Press (2008), are taken from four courses of lectures on the subject of negative dialectics. Presented while Adorno was writing the book *Negative Dialectics*, the lectures address the themes Adorno developed at the beginning of the book. Compared to the rather abstruse text and notoriously difficult (1973) English translation of *Negative Dialectics*, these lectures provide the English reader a relatively succinct explication of some of

Adorno's guiding theoretical motifs in his own (translated) words. We will make use of these lectures in this chapter where applicable. We have cross-referenced Adorno's (2008 [1965/66]) lectures with the English translation of *Negative Dialectics* (1973 [1966]) and have indicated this comparison below where necessary.

2   This is not to imply that Adorno was an orthodox Hegelian. Indeed, as J.M. Berstein (2004: 20) notes, the thought of anyone being an orthodox Hegelian is contradictory insofar as philosophy, for Hegel, is "one's own time and the history producing one's own time expressed in thought."

3   Brian O'Connor (2004, 2013) has argued that the most significant impact of Hegel's philosophy on Adorno is his theory of experience. This is substantiated by the following statement, taken from Adorno's (1963) book on Hegel: "These days it is hardly possible for a theoretical idea of any scope to do justice to the experience of consciousness, and in fact not only the experience of consciousness but the embodied experience of human beings, without having incorporated something of Hegel's philosophy" (Adorno, 1993 [1963]: 2). According to O'Connor (2004: 29), the introduction to Hegel's *Phenomenology* "provides Adorno with a model of how consciousness is determined through the experience of objects."

4   According to Adorno (2008: 25), "negativity of this kind is made concrete and goes beyond mere standpoint philosophy by confronting concepts with their objects and, conversely, objects with their concepts." See also Adorno (1973 [1966]: 4–6).

5   We do not engage the question of the degree to which Adorno's critique of Hegel is justified nor do we question the degree to which Adorno's critique rests on possible misunderstandings of Hegel's philosophy (on these and related issues, see Bernstein, 2004; Coyle, 2011; and O'Connor, 1999, 2004, 2013).

6   As Adorno remarks about the tasks of Marxism with respect to philosophy: "it confirms Hegelian idealism as prehistory's knowledge of its own identity. But it puts it back on its feet by unmasking that identity as prehistorical" (2003 [1942]: 95). Hegel's dialectic is thus unable to discern that "the law that (...) governs the restlessly destructive unfolding of the ever-new consists in the fact that at every moment the ever-new is also the old lying close at hand" (2003 [1942]: 95). The task of philosophy after Hegel is to "unmask the identity" of the "ever-new" as "prehistory" so that human beings can address history freely. As Adorno concludes "Only he who recognizes that the new is the same old thing will be of service to whatever is different" (2003 [1942]: 96).

7   Adorno is making reference to Marx's assertion that "philosophy can only be realized by the abolition [*Aufhebung*] of the proletariat, and the proletariat can only be abolished by the realization of philosophy" at the end of his introduction to his critique of Hegel's *Philosophy of Right* (1978

[1844]: 65). Significantly, Karl Korsch's *Marxism and Philosophy*, published the same year as Lukács' *HCC*, ends with this quote by Marx but prefaces it by writing: "[bourgeois] consciousness must be philosophically fought by the revolutionary materialistic dialectic, which is the philosophy of the working class." He concludes: "this struggle will only end when the whole of existing society and its economic basis have been totally overthrown in practice, and has been totally surpassed and abolished in theory" (Korsch (1970 [1923]: 85).

8 The model for the objective antagonistic world is, according to Adorno, "the fact that we live in an antagonistic society" (Adorno, 2008 [1965]: 8; cf. Adorno, 1973 [1966]: 151–153). As he explains, "[The] profit motive which divides society and potentially tears it apart is also the factor by means of which society reproduces its own existence (...) The factors that define reality as antagonistic are the same factors as those which constrain mind, i.e. the concept, and force it into its intrinsic contradictions. To put it in a nutshell, in both cases we are dealing with the principle of mastery, the mastery of nature, which spreads its influence, which continues in the mastery of men by other men and which finds its mental reflex in the principle of identity, by which I mean the intrinsic aspiration of all mind to turn every alterity that is introduced to it or that it encounters into something like itself and in this way to draw it into its own sphere of influence."

9 According to J.M. Bernstein's (2004: 32) interpretation of Adorno, "Logic and modern science are the quintessential expression of conceptual unification theoretically, while practically it is the relegation of use value to exchange value that performs the task." On the relationship between Adorno's theory and Marx's materialism, see Cook (2006, 2011), Jarvis (2004), and Stone (2006). On the relationship between Adorno and Marxism, see Cutrone (2013).

10 Rose (1976) is keen to point out that "totalizing" is not the same as "total." The identitarian logic cannot be total.

11 According to Tiedemann (2006: xvi), Adorno used the term spell to refer to the "eternal sameness of the historical process."

12 Although we shall make extensive use of examples from the American context, this is not meant to downplay the significance of global dynamics. Indeed, the totalizing nature of the administered society to which Adorno refers transcends national boundaries even as it is manifest differently in different societal contexts.

13 We return to specify the nature of this form of growth in our discussion of Postone in Chapter 3.

14 Christopher Lasch, writing in 1968, provides an excellent illustration of the character of the crisis in American liberalism in the 1960s: "After twenty years of the cold war, the focus of American politics has shifted far to the right. The liberal strategy of maintaining economic growth through

arms spending, of containing revolution through a series of limited police actions, and of buying off domestic discontent by building superhighways and cars by means of which the newly prosperous ethnic constituencies, still the backbone of the liberal-welfare coalition, could escape the cities to the consumer paradise of the suburbs—this strategy fell apart against the unexpected obstacles of Vietnam, ghetto riots, and student rebellion."

15 As we shall return to elaborate below, this subjectivity—articulated through the critique of affluence—is the foundation of the Anthropocene history.

16 In much the same way, consciousness around the 1952 smog did not end pollution, but rather shifted it spatially and qualitatively as the abatement of municipal pollution brought about new environmental problems (e.g., climate change).

17 We refer to this social conception of nature as contemporary ecological subjectivity in order to indicate the non-identity of the relationship between the "objective" dimension and the "subjective" dimension of the environment–society problematic; between social structure and forms of consciousness.

18 The contemporary earth sciences, for example, are a product of military patronage during the first decades of the Cold War (Doel, 2003).

19 On the environmental impact of the post-WWII constellation of power, see Hooks and Smith (2005, 2012); on the sociopsychological impact of this constellation, see, for example, Orr (2006).

3
Moishe Postone (1942–) and the Critique of Traditional Marxism: Helplessness and the Present Moment of the Great Acceleration

Abstract: *This chapter situates Moishe Postone's critique of traditional Marxism in relation to the present moment of the Great Acceleration. We engage a close reading of Postone's reinterpretation of Marx's mature theory of capital with specific focus on the linkage between economic growth and ecological degradation, and how this linkage is necessarily connected to social domination under modern capitalist society. Postone's Marxian theory is significant because, as we demonstrate, it allows one to grasp societally induced environmental degradation following WWII in a critical and reflexive manner. The chapter concludes by discussing the growing sense of helplessness that defines the present moment of the Great Acceleration.*

Stoner, Alexander M. and Andony Melathopoulos. *Freedom in the Anthropocene: Twentieth-Century Helplessness in the Face of Climate Change.* New York: Palgrave Macmillan, 2015. DOI: 10.1057/9781137503886.0008.

This chapter engages a close reading of Moishe Postone's reinterpretation of Marx's mature theory of capital. In doing so, we direct focus toward "the logic of modern capitalist society in general, and its early twenty-first century incarnation in particular" (Stoner, 2014: 622), in order to explicate the ways in which Postone's theory elucidates the present moment of the Great Acceleration. Postone's theory is important because, as we demonstrate later, it allows one to more fully understand the intricate linkage between economic growth and ecological degradation, and how this linkage is necessarily connected to social domination under modern capitalist society. Accordingly, we articulate Postone's critique as a critical and reflexive account of socioecological domination throughout the latter half of the twentieth century.

While both Lukács and Adorno engaged Marx's critique to develop their own systematic critiques of the effects of the capitalist mode of production on patterns of political, social, and cultural reproduction during the years following WWI (Lukács' critique of reification) and the first two decades following WWII (Adorno's critique of identity thinking), Postone's critique of "traditional Marxism"[1] attempts to grasp the interrelationship between the "inner logic" of capitalism at a later stage of development and its effects in determining patterns of political, social, and cultural reproduction.[2] In doing so, Postone advances a critical Marxian theory that grasps increasing productivity, technical development, and economic growth in relation to society as a whole and its ability to freely develop. Although economic growth and development have been routinely cited in explanations of the advent of both the Anthropocene and the Great Acceleration, they are often accorded an independent, causal existence; hence, the Great Acceleration is typically viewed as being driven exogenously to destruction. While Postone's theory is certainly better equipped to decipher the dynamics of capital at the beginning of the twenty-first century and, by extension, the present moment of the Great Acceleration, as we shall see, his theory also coincides with a world devoid of adequate political practice, in which revolutionary social theory exists without revolutionary practices.

1 Critique of traditional Marxism

Moishe Postone's reinterpretation of Marx's mature theory of capital, which he advances through a critique of traditional Marxism, must be viewed against the background of the historical failures of Marxism in order to adequately contextualize his work in relation to the present moment of the Great Acceleration within the Anthropocene. One effect of this failure has been an increased skepticism regarding the applicability of Marx's theory to the current historical moment, as expressed, for example, by theories of post-structuralism and deconstructionism which, as Murthy (2009: 9–10) explains, "seem to have the advantage of giving up totalizing narratives and grandiose projects of human emancipation." Indeed, the failure that attends the Great Acceleration appears to have rendered impossible the very idea that human beings might act as agents that can freely transform their history.[3] Here, we are once again confronted with the paradoxical nature of the environment–society problematic. In the present moment of the Great Acceleration we are faced with a situation in which increasingly sophisticated knowledge of objective biophysical threats is met by the constant reminder that society will breach any move toward freedom from heteronomy. Much as Marx's historical moment enabled him to critically reflect upon the fundamental nature and meaning of social transformation, Postone's critique of traditional Marxism grasps the (supposed) irrelevance of Marx's theory as the projection of a reified understanding of twentieth-century structural transformations (Postone, 2006). Without the ability to critically recognize the internal tensions of capital, and its current global neoliberal incarnation in particular, oppositional politics at the beginning of the twenty-first century find themselves confined to reactionary forms of "resistance," giving way to a profound sense of helplessness (Postone, 2006).

Although Marx's work provides a systematic critical theory of modern society, Postone contends that Marx's ideas have been appropriated through a traditional theoretical lens. In contrast to traditional Marxism, which affirms the centrality of labor in capitalist society, Postone argues that Marx's focus on the central role played by labor in capitalist society is fundamental to his *critical* theory of modern society. Marx's critical theory, according to Postone (1993: 307), conceptualizes labor as "a determinate mode of social mediation" that also structures a form of abstract domination unable to be grasped adequately when approached in traditional terms. Significantly, Postone's nuanced reading of Marx allows for an understanding of larger processes of social domination in light of both "state-centric capitalism" and "socialism" during the second and third quarters of the twentieth century as well as the most recent incarnation of capital, expressed in and accelerated through the advent of neoliberalism during the 1970s. Whereas Adorno waged his critique of identity within "administered" society whose regressive form of consciousness coincides with the full-scale development of the Great Acceleration, Postone's critique of traditional Marxism highlights the present moment of the Great Acceleration by specifying the current penetration of neoliberal global capitalism in and through the last gasp of its previous state-centric form, which not only accounts for the nonidentity of subject and object of practice (or, the theory/practice divide) as Adorno did, but also makes clear recognition of this deeper dynamic will not lead to a change automatically, but rather would require working through the deeper social structure of value, something, as Postone (2006) notes, has all but evaporated. As with the shift from Lukács to Adorno discussed previously, the shift from Adorno's critique of identity thinking (administered society and the emergence of consumerism in the culture industry) to Postone's critique of traditional Marxism (the rise of neoliberal global capitalism without the attendant growth of international Marxian politics) must be traced back into social structure so as to illuminate the historical meaning this shift registers.

Postone's reading of Marx's critical theory of modern society begins with the *Grundrisse rise der Kritik der politischen Ökonomie*, an unfinished manuscript written by Marx in 1857/58 though left unpublished until 1939 (and first published in English in 1973). The *Grundrisse*, according to Postone (1993: 21–22), indicates the categories of Marx's analysis are historically specific and that Marx's mature theory is not a critique of capitalism from the standpoint of labor but rather a critique of labor itself.

Postone advances this claim by focusing on one particular section of the *Grundrisse*, entitled "Contradiction between the foundation of bourgeois production (value as measure) and its development. Machines, etc." Marx (1974 [1857/58]: 704) begins this section with the following statement: "The exchange of living labour for objectified labour time—i.e. the positing of social labour in the form of the contradiction of capital and wage labour—is the ultimate development of the value-relation and of production resting on value." Postone (1993: 24) argues that the title and first sentence of this section indicate that "for Marx, the category of value expresses the basic relations of production of capitalism—those social relations that specifically characterize capitalism as a mode of social life—as well as that production in capitalism is based on value." In other words, value is not simply a market category, not simply a mechanism whereby equilibrium is achieved.⁴ Rather, value is both a *mediating* category and a category of *alienation*.⁵

That value is "both a determine form of social relations and a particular form of wealth" (Postone, 1993: 24) is the backbone of Postone's rereading of Marx. Postone (1993: 24) emphasizes that when Marx (1974 [1857/58]: 704) states, "the quantity of labour employed, as the determinant factor in the production of wealth," he indicates that what characterizes value as a determinate form of social relations and a particular form of wealth is that it is a social form that expresses and is based on the expenditure of direct labor time. As Marx (1974 [1857/58]: 704) continues, "But to the degree that large industry develops, the creation of real wealth comes to depend less on labour time and on the amount of labour employed than on the power of the agencies set in motion during labour time, whose 'powerful effectiveness' is itself in turn out of all proportion to the direct labour time spent on their production, but depends on the general state of science and on the progress of technology, or the application of this science to production."

It is in this passage that the contradiction between *value* and material *wealth*, which we return to elaborate later, is most apparent. The difference between material wealth and value, as Postone (1993: 25) explains, is that value is a form of wealth that depends on labor time and the amount of labor employed whereas material wealth does not. This implies that value is a historically specific form of wealth—it is not, in contrast to the traditional Marxist interpretation, a transhistorical form of wealth that could be distributed differently in different societies (Postone, 1993: 25). Similarly, the labor that constitutes value is not a property of labor in general but rather the historically specific temporal dimension of labor under capitalism that constitutes value as a form of wealth (Postone, 1993: 123). What is specific about capitalism—the social formation based on the commodity form—is, as Postone emphasizes, "*Personal independence* in the framework of a system of *objective* [*sachlicher*] *dependence*" (Marx, quoted in Postone, 1993: 125 [original emphases]). Citing Marx's *Grundrisse* (see Marx, 1974 [1857/58]: 164), Postone (1993: 125) stresses that this "'objective' dependence is social; it is 'nothing more than social relations which have become independent and now enter into opposition to the seemingly independent individuals; i.e., the reciprocal relations of production separated from and autonomous of individuals.'" As Marx (1974 [1857/58]: 164 [original emphasis]) notes, "individuals are now ruled by *abstractions*, whereas earlier they depended on one another." It is on this basis that value necessarily assumes a distinct, twofold phenomenal form: "the value of any given commodity is manifested first, independent of that commodity's use-value, and second, common to that commodity and all others" (Marx (1976 [1867]: 139). Commodity-producing labor is both particular (as concrete labor, a determinate activity that creates specific use-values) and socially general (as abstract labor, a means of acquiring the goods of others) (Postone, 1993: 151). It is this so-called double-character that, as a defining feature of labor under capitalism, expresses alienated social relations. As an expression of alienation (understood as a form of social mediation), commodity-determined labor consists of isolated individual labor while simultaneously assuming "the form of abstract generality" (Marx, quoted in Postone, 1993: 47). What makes labor general, and thus makes abstract labor *the* social mediation in capitalist society, is its social function. Postone (1993: 27) argues that in Marx's analysis of value "the basic social relations of capitalism, its form of wealth, and its material form of production" are interrelated.

Here Postone (1993: 27) stresses the significance Marx attaches to the increasingly anachronistic character of value as a measure of wealth vis-à-vis the immense wealth-producing potential of the industrial process of production:

> Labour no longer appears so much to be included within the production process; rather, the human being comes to relate more as watchman and regulator to the production process itself. (What holds for machinery holds likewise for the combination of human activities and the development of human intercourse.) No longer does the worker insert a modified natural thing [*Naturgegenstand*] as the middle link between the object [*Objekt*] and himself; rather, he inserts the process of nature, transformed into an industrial process, as a means between himself and inorganic nature, mastering it. He steps to the side of the production process instead of being its chief actor. In this transformation, it is neither the direct human labour he himself performs, nor the time during which he works, but rather the appropriation of his own general power, his understanding of nature and his mastery over it by virtue of his presence as a social body—it is, in a word, the development of the social individual which appears as the great foundation-stone of production and of wealth. The *theft of alien labour time, on which the present wealth is based,* appears a miserable foundation in face of this new one, created by large-scale industry itself, (Marx, 1974 [1857/56]: 705 [original emphases])

Gleaning insight from the passage earlier, Postone (1993: 24–25) argues that, for Marx, as capitalist industrial production develops value becomes an increasingly less adequate measure of material wealth. Marx (1974 [1857/58]: 705) does, of course, recognize that the persistence of value as a measure of material wealth produced, despite being increasingly anachronistic, remains a necessary structural precondition of capitalist society even though the potential embodied in the forces of production increasingly render production based on value obsolete (Postone, 1993: 25). Interpreting Marx in this way, Postone (1993: 27) claims that because production remains tied to value, where labor time is the sole measure of wealth, the abolition of value would signify the end of heteronomous (capitalist) labor.

Plausibly remedying our current ecological predicament (in which ecological degradation is compounded in proportion to our awareness of these problems) is not a matter of increasing technology or more rigorous scientific understandings of global climate change. Although science and technology provide important tools through which solutions to environmental problems might be pursued, in modern capitalist society science and technology are constituted in alienated form. Capital, as opposed to conscious and free individuals, dictates science and technology production. The environment–society problematic is not, then, a matter of science and technology per se but rather a matter of deep social history and politics. The contradiction between wealth and value (i.e., the increasingly anachronistic character of value as a measure of wealth) points toward capital's historically determinate socioecological domination as well as its possible supersession.

2   Nature of the contradiction: value and material wealth

In considering the contradiction between material wealth and value it is important to bear in mind (following Marx, 1976 [1867]: 126 as well as Postone, 1993: 194) that value and material wealth are two very different forms of wealth, which differ both qualitatively and quantitatively.[6] Whereas value is a function of the exchange value dimension of the commodity form, material wealth is a function of its use value dimension. The two dimensions are related through the commodity form of labor as a function of time.[7] More specifically, the magnitude of value is the objectification of what Marx refers to as *socially necessary labor time*.

Marx (1976 [1867]: 129) defines socially necessary labor time as follows: "the labour-time required to produce any use-value under the conditions of production normal for a given society and with the average degree of skill and intensity of labour prevalent in that society." But because this quantitative measure cannot be based on concrete labor alone, it must instead be based on abstract (i.e., socially necessary) labor.[8] However, if the magnitude of value depends on socially necessary labor time, then when average productivity is increased "the average number of commodities produced per unit time" is increased as well, which thereby decreases "the amount of socially necessary labor time required for the production of a single commodity and, hence, the value of each commodity" (Postone, 1993: 193).[9] The magnitude of total value produced, then, is a function only of the objectification of abstract labor time expenditure (measured in terms of constant time units) (Postone, 1993: 189).

Examining the contradiction between value and wealth we can see that "the magnitude of value is a function of the expenditure of abstract labor time," whereas "material wealth is measured in terms of the quantity and quality of products produced" (Postone, 1993: 193). Following Postone, the nature of the contradiction between material wealth and value can be elucidated by taking into account Marx's (1976 [1867]: 129) example of the introduction of power-loom weaving into the textile industry during the English Industrial Revolution:

> Let us assume that, before the power loom was introduced, the average hand-loom weaver produced 20 yards of cloth in one hour, yielding a value of x. When the power loom, which doubled productivity, was first introduced, most weaving was still done by hand. Consequently, the standard of value—socially necessary labor time—continued to be determined by hand-loom weaving; the norm remained 20 yards of cloth per hour. Hence, the 40 yards of cloth produced in one hour with the power loom had a value of $2x$. However, once the new mode of weaving became generalized, it gave rise to a new form of socially necessary labor time: the normative labor time for the production of 40 yards of cloth was reduced to an hour. Because the magnitude of value yielded is a function of (socially average) time expended, rather than the mass of goods produced, the value of the 40 yards of cloth produced in one hour with the power loom fell from $2x$ to x. Those weavers who continued to use the older method, now anachronistic, still produced 20 yards of cloth per hour but received only $½x$—the value of a socially normative half hour—for their individual hour of labor. (Postone, 1993: 288)

In fully developed capitalism (i.e., where relative surplus value[10] is the dominant form), increasing productivity (so as to yield a larger output per hour worked) is a primary means through which capitalists attempt to increase their profits. Yet, as indicated in the example earlier, this is only effective temporarily. Once a given level of productivity becomes generalized at the level of society as a whole, this then becomes the basis against which a new socially necessary labor hour is measured.[11] Postone explains that Marx's category of socially necessary labor time is something other and more than labor time expenditure. According to Postone (1993: 191), Marx's category of socially necessary labor time "expresses a general temporal norm resulting from the actions of the producers, to which they must conform." Socially necessary labor time has an additional social necessity because, in a society driven by the endless pursuit of profit, the social whole "is structured by value as the form of wealth and surplus value as the goal of production" (Postone, 1993: 302). Postone emphasizes that the reference point for socially necessary labor time, as the determination of a commodity's magnitude of value, is society as a whole: "Viewed from the perspective of society as a whole, the concrete labor of the individual is particular and is *part* of a qualitatively heterogeneous *whole*; as abstract labor, however, it is an individuated *moment* of a qualitatively homogenous, general social mediation constituting *a social totality*" (Postone, 1993: 152 [original emphases]). This is crucial in understanding the role time plays in the capitalist mode of production as an external necessity. That the reference point for socially necessary labor time is society as a whole implies "a tension and opposition between individual and society which points to a tendency toward the subsumption of the former by the latter" (Postone, 1993: 192). Socially necessary labor time therefore illustrates "a form of social life in which humans are dominated by their own labor and are compelled to maintain this domination" (Postone, 1993: 302). For Postone (1993: 287–288), Marx's example of the power loom "indicates that when the commodity is the general form of the product, the actions of individuals constitute an alienated totality that constrains and subsumes them."

Hence, our contention in Chapter 1 that the Anthropocene presupposes a historical context constituted by alienated social relationships. In this context, "people do not really control their own productive activity or what they produce but ultimately are dominated by the results of that activity" (Postone, 1993: 30).[12] Reworking Marx's (1976 [1867]: 137) position, Postone (1993: 195) stresses that while the forces of production of labor are analogous to the use-value dimension, productivity, as an expression of "the acquired productive abilities of humanity," is socially constituted in alienated form insofar as value is the dominant form of social wealth in capitalist society. Instead of being appropriated and controlled by people, the acquired productive abilities of humanity dominate and control people as an alienating force.[13] But as Postone is keen to point out, that individuals are mere organs of the whole in Marx's analysis should not be construed as a championing of the "organs" in opposition to the "whole"; Marx's analysis is a critique of the whole itself. Marx's critique is rooted in the contradiction between wealth and value: increases in productivity do *not* correspond to growing amounts of value per unit time; products function simply as material "bearers of objectified time" (Postone, 1993: 312). We will expand on what this means shortly.

3 Socioecological domination and the production of value

As Marx discusses in the first chapter of *Capital, Volume 1*, a product becomes a commodity when it is transferred to another person via the medium of exchange. The incommensurability of commodities (as use-values) must somehow be rendered commensurable in order for an exchange to happen and this commensurability requires an objective measure of comparison, which Marx reduces to value. A commodity therefore appears to be the exchange of external objects mediated by the market. That the commodity is something other and more than that indicated by market exchange only becomes apparent retrospectively, particularly in the latter half of *Capital, Volume 1*, where Marx (1976 [1876]: 207–210) outlines how the true nature of money in capitalism may be concealed (cf. Postone, 1993: 265).

According to Postone (1993: 265), it is here in the latter half of *Capital, Volume I*, and the transition from the analysis of money to the category of capital, in particular, where "Marx unfolds a dialectical reversal in his treatment of money: it is a social means that becomes an end." Far from rendering commodities commensurable, money is a necessary form of appearance of their commensurability. As Postone explains:

> [B]ecause the circulation of commodities is effected by the externalization of [the] double character of [commodity-mediated labor]—in the form of

money and commodities—they seem to be mere "thingly" objects, goods circulated by money rather than self-mediating objects, objectified social mediations. Thus, the peculiar nature of social mediation in capitalism gives rise to an antinomy—so characteristic of modern Western worldviews— between a "secularized," "thingly" concrete dimension and a purely abstract dimension, whereby the socially constituted character of both dimensions, as well as their intrinsic relation, is veiled. (Postone, 1993: 264–265)

Postone's gloss (1993: 265) on Marx's discussion of money is in line with his critique of the way traditional Marxism misrecognizes labor: money, as the externalized expression of "abstract labor objectified as value" is "an externalized expression of the form of social mediation that constitutes capitalist society." Consequently, money appears to be natural (i.e., nonsocial). However, it is important to bear in mind, as Postone (1993) indicates in the passage earlier, that the distinction between the use-value dimension and the value dimension of the commodity form of social relations is not evident at the level of immediate experience.[14]

The commodity both has a form and is a form. Whereas the content of commodity's value dimension form is a social relation, the product (commodity) is brought into being through objectifying activity (labor). As a social form, the commodity has content—namely, abstract labor— and it is this latter, substantive dimension that underlies the formalism of the capitalist system which allows Marx to put forth his formalistic account of capital. Postone (1993: 268 [original emphases]) stresses that M-C-M'[15] is not the formula for a process whereby "*wealth* in general is increased"; rather, it is the formula for a process whereby "*value* is increased":[16]

> With capital, the transformation of (the commodity) form becomes an end and (…) the transformation of matter becomes the means to this end. Production, as a social process of the transformation of matter which mediates humans and nature, becomes subsumed under the social form constituted by labor's socially mediating function in capitalism. (Postone, 1993: 267)

According to Postone (1993: 267), the formula M-C-M' implies an immanent dynamic wherein "a quasi-natural network of social connections develops" which, "although constituted by human agents, lies beyond their control" (Postone, 1993: 264). With the production of value concrete human–ecological transformation via labor is subsumed under the form of abstraction, which is why the difference between M and M', which Marx calls *surplus value*, is "necessarily only quantitative" (Postone, 1993:

267).[17] Marx's concept of *capital as self-valorizing value* is an attempt to grasp this built-in dynamic to accumulate *ad infinitum*, which in fully developed capitalism, is marked not only by increasing productivity (so as to yield a larger output per hour worked) but as we shall see, a tendency for the rate at which productivity increases to accelerate over time.

Postone explains the development of relative surplus value, as a self-valorizing value, and its expansion (required by capital) as follows:

> With the development of relative surplus value, then, the directional motion that characterizes capital as self-valorizing value becomes tied to ongoing changes in productivity. An immanent dynamic of capitalism emerges, a ceaseless expansion grounded in a determinate relationship between the growth of productivity and the growth of the value form of the surplus. (Postone, 1993: 283)

Advances in productivity, as exemplified by the "runaway" growth that characterizes the Great Acceleration, "do *not* increase the amount of value yielded per unit of time, but they *do* increase greatly the amount of material wealth produced" (Postone, 1993: 197 [original emphases]).[18]

This means that although in fully developed capitalism, relative surplus value is acquired by way of increasing levels of productivity, and although increases in productivity turn out greater quantities of material wealth and reduce socially necessary labor time, these developments do not change the total value produced per abstract time unit (i.e., labor expenditure as measured by the independent variable, abstract time) because the "constant" time unit itself is a dependent variable, whose determination is dictated by productivity as a function of the use-value dimension of commodity-determined labor. The magnitude of total value produced depends only on the amount of abstract human labor time expended.

According to Postone (1993: 287–288), because the magnitude of value is determined by socially necessary labor time as the objectification of abstract labor expenditure at the level of society as a whole, once an increase in productivity becomes generalized the magnitude of value falls back to its previous level. The insidiousness of this dialectic of labor and time, according to Postone, is continuously enacted by way of a particular "treadmill effect." As he explains,

> The more closely the amount of surplus value yielded approaches the limit of the total value produced per unit time, the more difficult it becomes to

further decrease necessary labor time by means of increased productivity and, thereby, to increase surplus value. This, however, means that the higher the general level of surplus labor time and, relatedly, of productivity, the more productivity must be further increased in order to achieve a determinate increase in the mass of surplus value per determinate portion of capital. (Postone, 1993: 310–311)

Recall that abstract time, by definition, is measured in terms of constant time units, whereas productivity corresponds to labor's use-value dimension—value remains a form of social necessity despite the fact that its determination (abstract labor time expended) operates independent of changes in productivity. Hence, value becomes increasingly anachronistic as a form of social wealth in the face of the immense wealth-producing potential of modern industry.

Postone incorporates his analysis of the commodification of time (see footnote 7) in order to further elaborate this treadmill dynamic, whose initial determination "delineates the form growth *must* take in the context of labor-mediated social relations" (1993: 290 [original emphasis]). In doing so, Postone (1993) unfolds the category of time as a commodity whose abstract and concrete dimensions are synchronous with the intrinsic interaction between concrete and abstract labor. On the basis of this dialectic of labor and time, Postone is able to indicate how the duality of these social forms interact to transform and reconstitute capital's social totality.

Before examining further the temporal dimension of the necessity of value, it behooves us at this point to outline some of the implications of the foregoing discussion. First and foremost, it should be noted that Postone's reinterpretation of Marx allows for an explanation of the Great Acceleration in terms of the "treadmill dynamic" alluded to earlier—that is, in terms of a dialectic between quality/quantity and labor/time. The Great Acceleration could then be reconceptualized as the unfolding of the contradiction between wealth and value throughout the latter half of the twentieth century. This development indicates a historically specific pattern of "progress," which, drawing on Postone's reinterpretation of Marx, we specify later in terms of the growing necessity of accelerating the rate at which biophysical throughout increases. On the basis of Postone's appropriation of Marx's category of socially necessary labor time, we can understand how capital (as self-valorizing value) works *through* people, propelling forward a directional dynamic that can be circumscribed by the tendency to produce more and more in less and less time *ad infinitum*. Examining the Great Acceleration in this manner would allow for a critical, dynamic, and reflexive account of so-called economic growth[19]—that is, the form growth must take via the production of value—in terms of the inextricable connection between the intensifying domination of people by time, on the one hand, and the necessity of increasing biophysical throughput exponentially, on the other. Such an account of historical transformation is significantly different from that of the Anthropocene, which tends to view the Great Acceleration as being driven exogenously to destruction. Significantly, Postone's theory does not rely on causes outside society nor does it look to an external impending catastrophe to generate freedom and agency. Rather, Postone attempts to understand these developments in relation to society as a whole and its ability to freely develop. Within this critical theoretical framework, both history and freedom are grasped in terms of the double-sided nature of capital's social forms. In the following section we turn to examine why, despite being rendered increasingly anachronistic, the value form remains necessary.

4 Value as the continual necessity of the present

The concept of capital as self-valorizing value is significant for our purposes in examining the "runaway" character of the Great Acceleration because it specifies the form economic growth must take in this context. In fully developed capitalism, economic growth assumes a form in which both people and biophysical nature are increasingly rendered material "bearers of objectified time" (Postone, 1993: 312) at an accelerating rate.[20] And while capital appears to serve human ends, it is actually humanity which serves capital.[21] As Postone explains,

> This pattern of growth is double-sided for Marx: it involves the constant expansion of human productive abilities, yet tied as it is to an alienated dynamic social structure, this expansion has an accelerating, boundless, runaway form over which people have no control (...) [O]ne particular consequence implied by this particular dynamic—which yields increases in material wealth greater than those in surplus value—is the accelerating destruction of the natural environment. (Postone, 1993: 311)

In addition to the material wealth dimension of so-called economic growth grasped in quantitative terms (i.e., the tendency of the ratio of products produced per unit of labor to increase over time, which also implies increases in biophysical throughout), Postone's critical Marxian theory captures a dynamic whereby the referent against which such changes are measured undergoes a qualitative social transformation. Later we follow Postone in explicating how, as self-valorizing value, capital drives the production of value forward in time, continuously enacting its necessity in the present.[22]

Postone distinguishes between abstract and historical time and indicates a dialectic between the two. Historical time is analogous to the use-value dimension and may be considered a form of concrete time as constituted in capitalism. Historical time, according to Postone, "is the movement *of time*, as opposed to the movement *in time*" (1993: 294 [original emphases]). As Postone (1993: 294) explains, "The social totality's dynamic expressed by historical time is a constituted and constituting process of social development and transformation that is directional and whose flow, ultimately rooted in the duality of the social relations mediated by labor, is a function of social practice." Abstract time, on the other hand, is a function of the measure of value. In contrast to historical time, Postone points out that "Although the measure of value is time, the totalizing mediation expressed by 'socially necessary labor time' is not a movement *of time* but a metamorphosis of substantial time into abstract time *in space*, as it were, from the particular to the general and back" (1993: 293 [original emphases]).

With regard to the temporal dimension of the production of value, Postone (1993: 293) indicates that although the abstract temporal measure of value remains constant, "both the social labor hour and the base level of productivity are moved 'forward in time.'" This "substantive redetermination of the abstract temporal constant" necessitates accelerating the rate at which productivity increases (Postone, 1993: 292). As Postone (1993: 292) explains, in this sense the constant hour becomes "denser" as the amount of products produced increases. This "substantive redetermination" is not immediate, however, and is therefore not apparent at the level of appearances despite the fact that a substantive redetermination (as indicated in increases in productivity) has actually occurred (Postone, 1993: 292). The process through which the hour becomes "denser" cannot be expressed in abstract time because, as Postone (1993: 292) explains, the social labor hour (abstract time), although redetermined, is the "form against which change is measured." According to Postone (1993: 292–293), "The entire abstract temporal axis, or frame of reference, is moved with each socially general increase in productivity."

Because productivity is rooted in the use-value dimension of labor, it is possible to conceive of the "forward" movement of the abstract temporal frame of reference "as a mode of concrete time" (Postone, 1993: 293). That the interaction between capital's use-value and value dimensions can be conceived in this way is itself indicative of capitalism. As Postone (1993: 293 [original emphases]) explicates, the interplay of abstract labor and concrete labor sheds light on the foundation of Marx's analysis of capital, especially the fact that *"a feature of capitalism is a mode of (concrete) time that expresses the motion of (abstract) time."* Historical time within capitalist society, then, is socially constituted (via praxis) insofar as it is mediated by value. Within the critical Marxian framework advanced by Postone, human agency is always constrained, which is to say that social structure is not the opposite of agency but is rather constitutive of agency (cf. Lukács, 1923).

As mentioned previously, Postone analyzes the dialectic interplay between capital's dual forms as giving rise to an immanent dynamic—a dialectic of transformation and reconstitution between the abstract and concrete dimensions. This dialectic of transformation and reconstitution, according to Postone (1993: 294), is also operative between two forms of social necessity: (1) ongoing surface-level transformations and (2) the continuing reconstitution of the underlying conditions necessary for the production of value.[23] The dialectic of transformation and reconstitution implies that, as Postone (1993: 295) explains, "the Marxian analysis elucidates and grounds socially the historically dynamic character of capitalist society in terms of a dialectic of abstract and historical time."

While both forms of time are intrinsically related, the abstract temporal unit is distinct in that "it does not manifest its historical redetermination—it retains its constant form as *present time*" (Postone, 1993: 295 [original emphases]). Like the commodity form, the "social 'content' of the abstract temporal unit remains hidden" (Postone, 1993: 295). Moreover, value, as an expression of time as the present, represents an external social norm (Postone, 1993: 295).

Here we return again to Marx's example of the power loom, where Postone (1993) applies this insight as follows:

> The social labor hour in which the production of 20 yards of cloth yields a total value of x is the abstract temporal equivalent of the social labor hour in which the production of 40 yards of cloth yields a total value of x: they are equal units of abstract time and, as normative, determine a constant magnitude of value. Assuredly, there is a concrete difference between the two,

which results from the historical development of productivity; such a historical development, however, redetermines the criteria of what constitutes a social labor hour, and is not reflected in the hour itself. In this sense, then, *value is an expression of time as the present*. It is a measure of, and compelling norm for, the expenditure of immediate labor time regardless of the historical level of productivity. (Postone, 1993: 296 [original emphases])

Insofar as substantive changes effected by the use-value dimension, including, for example, societally induced environmental degradation, remain nonmanifest in terms of the abstract temporal frame of value, these changes cannot be recognized within the framework of the present.[24] Recall that the unfolding of capital's historical dynamic implies that as productivity increases value as a form of social wealth becomes increasingly anachronistic. The dynamic of capital gives rise to an ever-increasing disparity between the accumulated historical potential of scientific knowledge and the production of value. As Postone (1993) explains,

The dynamic of capitalism, as grasped by Marx's categories, is such that with this accumulation of historical time, a growing disparity separates the conditions for the production of material wealth from those for the generation of value. Considered in terms of the use value dimension of labor (that is, in terms of the creation of material wealth), production becomes ever less a process of materially objectifying the skills and knowledge of the individual producers or even the class immediately involves; instead, it becomes ever more an objectification of the accumulated collective knowledge of the species, of humanity—which, as a general category, is itself constituted with the accumulation of historical time. In terms of the use value dimension, then, as capitalism develops fully, production increasingly becomes a process of the objectification of historical time rather than of immediate labor time. According to Marx, though, value, necessarily remains an expression of the latter objectification. (Postone, 1993: 298)

The growing disparity between the accumulated historical potential of humanity and the production of value, however, does not automatically undermine the necessity represented by value; that is, the necessity of the present (Postone, 1993: 299), but rather changes the "concrete presuppositions of that present, thereby constituting its necessity anew" (Postone, 1993: 299). Value, as an expression of time, indicates that capitalism is simultaneously dynamic and static: "It entails ever-rising levels of productivity, yet the value frame of reference is perpetually reconstituted anew" (Postone, 1993: 299). Historical time, as constituted by capital's dynamic totality, is invariably naturalized "into the framework of the present, thereby reinforcing that present" (Postone, 1993: 300).

The expansion of surplus value necessitated by capital thus illustrates a specific proneness to accelerate the rate at which productivity increases which in turn necessitates accelerating the rate at which quantities of biophysical throughput increase. Again, "throughput"—that is, the rate at which the capitalist system achieves its goal of expanding surplus value—must be understood in reference to Postone's categorial appropriation of Marx's category of socially necessary labor time. As Postone explains,

> Capital produces material wealth as a means of creating value. Hence it consumes material nature not only as the stuff of material wealth but also as a means of fueling its own self-expansion—that is, as a means of effecting the extraction and absorption of as much surplus labor time from the working population as possible. Ever-increasing amounts of raw materials must be consumed even though the result is not a corresponding increase in the social form of surplus wealth (surplus value). The relation of human and nature mediated by labor becomes a one-way process of consumption, rather than a cyclical interaction. It acquires the form of an accelerating transformation of qualitatively particular raw materials into "matter," into qualitatively homogenous bearers of objectified time. (Postone, 1993: 312)

On this basis, the expansion of surplus value is a process whereby both people and biophysical nature increasingly become material "bearers of objectified time" at an accelerating rate. Examining the Great Acceleration in these terms allows one to more fully comprehend the linkage between economic growth and environmental degradation, and the ways in which this linkage is connected to social domination under modern capitalist society.

5 Helplessness

Postone's critique of traditional Marxism is significant in light of the Great Acceleration because it provides an understanding of how one might plausibly move beyond this context while being a part of it—something that remains elusive if we conceptualize this history in terms of the Anthropocene. The potential to move beyond the present moment of the Great Acceleration is inherent in the contradictions of capital; specifically, "between the actuality of the form of production constituted by value, and its potential" (Postone, 1993: 28). Recognition of this dynamic does not, however, lead to a change automatically, but rather requires working through the deeper social structure of value, something, as Postone (2006: 95) notes, has all but evaporated: "Although indeterminate, a postcapitalist social form of life could arise only as a historically determinate possibility generated by the internal tensions of capital, not as a 'tiger's leap' out of history." While the necessity of value *ought* to be the manner in which one could develop the possibility for the overcoming of capitalism, the fact that it does not and only generates "resistance"[25] instead is indicative of a profound sense of helplessness regarding the capacity of society to self-consciously transform itself in ways that are not predetermined from the outset (Postone, 2006). That the increasing penetration of the value form (and with it the potential for triggering a social political crisis) has given way to resistance is the measure of this helplessness.

Interestingly, it is in an earlier article that Postone (1978) recovers Marx through the social movements of the 1960s (cf. Cutrone, 2014). If Marx could use the critique of political economy to discern the potential of workers to attain consciousness of the contradictory totality, Postone suggests that some of the persisting features of the failure of environmentalism (e.g., the implicit questioning of labor and work in the late 1960s and early 1970s) are an indication of the potential of a properly constituted ecological politics to provoke a social crisis. Here Postone detects how something like contemporary ecological subjectivity might provoke a crisis, and he does so in a way reminiscent of Marx's recognition that the crisis provoked by proletarian politics through Bonapartism evinced the possibility of overcoming the deeper structure of society. As mentioned, Postone's approach is rooted in the Marxian theory of praxis as a form of social mediation, which he contends implies that Marx's critical theory is rooted in a critique of labor as self-generated domination. In this sense, Postone (1978) suggests that the social movements of the 1960s, insofar as they express the desire to move beyond wage-labor, might be seen as an instantiation of what he refers to as "class transcending consciousness," which calls into question immediate productive labor itself. By focusing on the "*contents* of needs and consciousness" (Postone, 1978: 783 [original emphasis])—for example, struggles over the working day versus struggles against the harmful effects of pesticides—Postone further distinguishes between the "historical *possibility*" of socialism and the "*probability* of revolution" and suggests that the distinction between the two can be conceptualized as occupying different axes of historical time (Postone, 1978: 783 [original emphases]). Within this framework, the problem of the possibility of socialism (and with it the possibility of freedom) is clarified horizontally, as it were, "as the historical changing content, independent of degree of militance" (Postone, 1978: 783). The problem of the probability of revolution, on the other hand, is clarified vertically, "moving from an abstract analysis of the metahistory of the social formation to a consideration of more immediate, concrete, and contingent factors" (Postone, 1978: 783).

As we have seen, Postone reconceptualizes the contradiction between the forces of production and the relations of production as a contradiction "between the actuality of the form of production constituted by value, and its potential" (Postone, 1993: 28). Applied to the relationship between objective social structure, contemporary ecological subjectivity as an instantiation of class-transcending consciousness might then be conceptualized as an attempt to mediate "social objectivity in which a certain structure of labor has become anachronistic (…) even when this experience is not politically articulated" (Postone, 1978: 784). Indeed, the relationship between contemporary ecological subjectivity and "historically emergent contradictions of the social totality" (Postone, 1978: 785) is far from linear or direct.

Gleaning insight from Postone, we can return to the failure of contemporary environmentalism discussed in the previous chapter. Although the early environmentalist criticism was articulated at a time when the material expansion of the post-WWII regime had developed to such an extent that it became possible to question the necessity of wage labor,[26] the rise of contemporary environmentalism did not correspond to a related shift in how society was organized. In fact, the exact opposite occurred as the condition of full employment in the 1960s and early 1970s gave way to high unemployment as the 1970s wore on. Moreover, the growth of environmentalism throughout the 1970s and 1980s corresponded with the advent and continuation of neoliberalism, establishing a new round of capital accumulation. The problem of identifying capitalism with one dimension (e.g., employment, distribution, technology) is precisely the problem Postone is warning against. This is something the environmental movement was unable to understand as their political leverage, built up over the early 1970s in the United States (e.g., Nixon and Clean Air Act), was readily undermined and evaporated through a new set of social discontents and the onset of high unemployment. Consequently, environmental discontents never become adequately objectified and ended up, for example, being readily incorporated first into Nixon's environmental legislation and then into green entrepreneurship after environmental "militancy" began to marginalize the movement from society. This is not to suggest a static opposition between "militancy" and entrepreneurship, to be sure. Rather, we must regard both as integral moments in the disintegration of "state capitalism."

On this basis, we contend that contemporary ecological subjectivity as an instantiation of class-transcending consciousness is grounded in the socioecological tensions underlying the production of value, which became increasingly exacerbated throughout the latter half of the twentieth century. As such, it is part and parcel of the very process of compounding human impact on the environment, which it nonetheless seeks to move beyond. Yet, recognition of this growing tension (between the necessity of value and the need and desire to move beyond the value form) has yet to occur on a level adequate to the scale of the problem at hand. Postone's critique of traditional Marxism, through which he explicates the production of value in relation to the tension between freedom and necessity, is significant in this regard because, as we have seen, it sheds light on the continual necessity of the Great Acceleration while specifying the possibility of a different form of social relations. At the same time, Postone's insight must be understood in relation to the unfolding of the latter half of the twentieth century—a period characterized by the growing inability to connect the theory (of totality) with the "forms of appearance" (discontents) politically, which is something that Lukács, writing during the first quarter of the twentieth century, could take for granted. Moreover, the social crisis provoked by the discontents in the 1960s was not on a scale (i.e., was not sustained to render political and hence conscious) of those arising from the revolutionary tide of 1917–1919. Postone registers the potential crisis of "state capitalism" in the new forms of discontent in the 1960s and he identifies in these discontents the first instance of "class-transcending" consciousness. But this only shows the New Left's lack of understanding of what the crisis of "state capitalism" was (Cutrone, 2014). The New Left obscures the disintegration of "state capitalism" without being able to advance it politically (i.e., consciously).

Similarly, contemporary environmentalism is a form of discontent with "state capitalism." It registers the crisis, but in ways that readily gives expression to the transformation into neoliberalism. This is evident not only in "green" consumerism but also in the aggressive anti-technology stance (i.e., against the form technology took under "state capitalism"). Of course, because environmentalism is unable to understand itself as a product of "state capitalist" science (as discussed in the previous chapter), it is also unable to comprehend how new technologies (e.g., GMOs and next-generation pesticides) are also a product of its discontents. The inability of environmentalism to recognize itself as the product of that which it seeks to move beyond is the point at which Postone's critique of traditional Marxism registers the present moment of the Great Acceleration. Hence, the problem of being able to discern the absence of political forms adequate to addressing runaway development, but without any practical recourse. In this sense, the Anthropocene, which warns against an impending "outside" ecological threat, is true. But because the Anthropocene obscures the meaning of this history, it unintentionally projects the helplessness of the present back over the history of modern society. It is not possible to conceive of the Anthropocene independent of history and freedom, regardless of the current proposal to do so.

Notes

1. Postone uses the term "traditional Marxism" to cover an array of approaches and theorists working within the Marxist tradition. "Traditional Marxism" includes not only Ricardian Marxism, but also the Marxism of Lukács and members of the Frankfurt School, including Adorno. Discussing the critical theory of Lukács and the Frankfurt School, Postone (1993: 84–85) notes, "These attempts considerably broadened and deepened the scope of social critique and called into question the adequacy of traditional Marxism as a critique postliberal modern society. Yet, in seeking to formulate a more adequate critique, Critical Theory ran into serious theoretical difficulties and dilemmas. These became manifest in a theoretical turn taken in the 1930s, wherein postliberal capitalism came to be conceived as a completely administered, integrated, one-dimensional society, one that no longer gives rise to any immanent possibility of social emancipation."
2. This point is developed by Dahms (2011), who provides a critical analysis of the early Frankfurt School critique of capitalism and Postone's related contributions.
3. Although post-humanism appears to diverge from the relative optimism of liberal environmentalism by emphasizing the ineffectual nature of political activity, both positions have common ground in the standpoint of the immediate present where they entertain, as opposed to critically reflect on, this moment.

4 The traditional Marxist interpretation of the theory of surplus value, codified in introductory texts to Marx's theory, interprets the theory of surplus value by emphasizing the source of profit-making in labor exploitation where the value produced by labor in excess of the value of labor power goes to the capitalist class as profit.

5 To reiterate a point made in the Introduction, alienation, as a form of social mediation, is a process of self-generated domination in which, through concrete forms of social practice, humans create structures that in turn dominate them (Postone, 1993). As such, alienation is a dynamic mediating process between the subject–object dimensions of the environment–society problematic (see, e.g., Stoner, 2014).

6 Material wealth, as Postone (1993: 194; see also Marx, 1976 [1867]: 134, 136–137) explains, "arises from the interactions of humans and nature, as mediated by useful labor (…) its measure is a function of the quantity and quality of what is objectified by concrete labor, rather than the temporal expenditure of direct human labor." Value, however, is quite different. Unlike material wealth, "nature does not enter directly into value's constitution" (Postone, 1993: 194). As Postone (1993: 195; see also Marx, 1976 [1867]: 137) notes, the magnitude of value is "not a direct expression of the quantity of products created or of the power of natural forces harnessed; it is, rather, a function only of abstract time."

7 Postone (1993: 200–216) discusses the commodification of time—a category which he appropriates, following Marx, in terms of the double-sided nature of capital's social forms. Thus, Postone distinguishes between: (1) concrete time (time as a dependent variable) and (2) abstract time (time as an independent variable). Whereas *concrete time*, which, as Postone (1993: 201) explains, dominated conceptions of time before the rise of modern capitalist society in Western Europe, functioned as a dependent variable, "Abstract time is an independent variable; it constitutes an independent framework within which motions, events, and action occur. Such time is divisible into equal, constant, nonqualitative units" (Postone, 1993: 202). Postone traces the historical development of abstract time, which, as he (Postone, 1993: 202) shows, coincides with the spread of the commodity form of social relations.

8 Following Marx (1976 [1867]), concrete labor, embodied in different use-values produced, refers to qualitative differences. As such, concrete labor and the use-value it produces do not provide an objective basis for the exchange of commodities. In order to be exchanged the incommensurability of commodities (as use-values) must be rendered commensurable. Commensurability requires an objective measure of comparison, and the property that allows commodities to be compared is what Marx calls value. Exchange-value therefore expresses value as that which transcends differences in commodities as use-values. Abstracted from the utilities

(use-value) of commodities, the only common characteristic left, according to Marx, is labor. And although value is simply expended labor (Marx, 1976 [1867]: 135), in order to establish what is common to all commodity-producing labor, Marx argues that it is still necessary to abstract from the qualitative differences of the concrete labor embodied in the use-value dimension of the commodity. This dimension of commodity-producing labor is designated by Marx's category "abstract labor."

9 As Marx (1976 [1867]: 131) notes, "The value of a commodity (...) varies directly as the quantity, and inversely as the productivity, of the labour which finds its realization within the commodity."

10 "I call that surplus-value which is produced by the lengthening of the working day, *absolute surplus-value*. In contrast to this, I call that surplus-value which arises from the curtailment of the necessary labour-time, and from the corresponding alteration in the respective lengths of the two components of the working day, *relative surplus-value*" (Marx, 1976 [1867]: 432 [original emphases]). The move toward relative surplus value is historically specific and refers to a context in which limits to the working day are given.

11 The full elaboration of this dynamic, however, will only become apparent in the following section after we have incorporated Marx's category of capital.

12 Similarly, Postone (1993: 31) recasts Marx's (1988 [1844]) earlier writings on alienation: "The theory of alienation implied by Marx's mature critical theory does not refer to the estrangement of what had previously existed as a property of the workers (and should, therefore, by reclaimed by them); rather, it refers to a process of the historical constitution of social powers and knowledge that cannot be understood with reference to the immediate powers and skills of the proletariat. With his category of capital, Marx analyzed how these social powers and knowledge are constituted in objectified forms that become quasi-independent of, and exert a form of abstract domination over, the individuals who constitute them."

13 Following Postone (1993: 195), and as we elaborate later, these productive abilities also become attributes of capital.

14 The elusively dynamic nature of the commodity form of social relations is evident in the distinction between values and prices, an issue that has been a source of fundamental confusion among the majority of Marxist economists. Postone (1993: 196) suggests that in order to flesh this out, one would have to "elucidate how a categorial distinction—such as that between value and material wealth—is indeed operative socially, although the actors may be unaware of it. One would need to show how people, acting on the basis of forms of appearance that disguise the underlying essential structures of capitalism, reconstitute these underlying structures. Such an exposition would need to show how these structures, as mediated by their forms of appearance, not only constitute practices that are socially constituting,

15 M-C-M (Money-Commodity-Money) is Marx's formula for "the transformation of money into commodities, and the re-conversion of commodities into money: buying in order to sell" (1976 [1867]: 248). Marx employs the formula M-C-M′—where the difference between M and M′ is necessarily only quantitative; that is, "buying in order to sell dearer" (1976 [1867]: 256)—to capture "the general formula for capital, in the form in which it *appears* directly in the sphere of circulation" (1976 [1867]: 257 [emphasis added]). According to Marx (1976 [1867]: 188), "Money as a measure of value is the necessary form of appearance of the measure of value which is immanent in commodities, namely, labour-time."

16 This appears to be a point where one might pursue an important comparison between the socioecological implications of Postone's critical Marxian theory, on the one hand, and Allan Schnaiberg's (1980) treadmill of production, on the other. Whereas Schnaiberg stresses the production of *wealth*, the treadmill dynamic Postone identifies emphasizes the production of *value*. Unfortunately, the detailed attention necessary for such an examination is well beyond the confines of the present book.

17 As we return to elaborate below, this implies that Marx's category "capital" is the dynamic unfolding of the wealth-value contradiction: "Capital itself is the moving contradiction, [in] that it presses to reduce labour time to a minimum, while it posits labour time, on the other side, as sole measure and source of wealth" (Marx, 1974 [1857/58]: 706).

18 It is important to be aware that the peculiarity of this dynamic, which can only be explained in relation to the temporal dimension of value, emerges only when relative surplus value becomes the dominant form of wealth (as it did during the nineteenth century). In other words, the tendency toward accelerating rates of increases in productivity (and therefore accelerating rates of increases in biophysical throughput) becomes apparent only when the working day has been limited.

19 As alluded to above, Steffen et al. (2011) accord causal significance to "economic growth" in explaining the advent of both the Anthropocene and the Great Acceleration, though they never define the term nor do they inquire into the context to which it refers.

20 "Capital produces material wealth as a means of creating value. Hence, it consumes material nature not only as the stuff of material wealth but also as a means of fueling its own self-expansion—that is, as a means of effecting the extraction and absorption of as much surplus labor time from the working population as possible" (Postone, 1993: 312).

21 "The objective content of (...) the valorization process is his [the capitalist's] subjective purpose, and it is only in so far as the appropriation of ever more

wealth in the abstract is the sole driving force behind his operations that he functions as a capitalist; i.e., as capital personified and endowed with consciousness and will" (Marx, 1976 [1867]: 254).

22 Although the process whereby a "new" labor hour becomes socially necessary entails a quantitative change, the nature of this dynamic cannot be grasped in purely analytic terms. The social necessity of socially necessary labor time is, in part, an external social necessity, insofar as it produces and is produced by alienated, heteronomous labor—that is, how our world is created in modern capitalist society.

23 Postone (1993: 294) indicates some of the basic determinations of the historical process specific to capital's dynamic totality (as analyzed by Marx). These include but are certainly not limited to: (1) the continual development of productivity, which entails (2) "massive transformations in the mode of social life of the majority of the population," and (3) "the constitution, spread, and ongoing transformation of historically determinate forms of subjectivity, interactions, and social values."

24 This raises the question of the possibility of critique. Here it is important to note that, as Postone (1993: 295) emphasizes, both abstract time and historical time are expressions of alienated social relations, which is why the opposition between these two moments is not an opposition between capitalist and noncapitalist moments. Similarly, the opposition between "growth" and "degrowth," which typifies environmental critique, does not point beyond, but rather is intrinsic to capital (see Postone, 1993: 317). Moreover, as Postone (1993: 17–18) notes, Marx's position "neither affirms the existence of a transhistorical logic of history nor denies the existence of any sort of historical logic. Instead, it treats such a logic as characteristic of capitalist society which can be, and has been, projected onto all of human history." According to Postone (1993: 18), it is in this way that Marx's theory "reflexively attempts to render plausible its own categories (…) Theory, then, is treated as part of the social reality in which it exists."

25 As in, for example, the anti-imperialism following 2001: "At the heart of this neo-anti-imperialism is a fetishistic understanding of global development—that is, a concretist understanding of abstract historical processes in political and agentive terms" (Postone, 2006: 96).

26 During the 1960s, levels of average unemployment fell below three percent in most OECD countries (Nickell et al., 2005), which increased the political leverage of the working class. Yet, we must admit that Adorno's (2003 [1969]: 118) warning against Keynesian triumphalism represents a more accurate description of this state of affairs: "the triumph of technical productivity deludes us into believing that utopia, which is irreconcilable with the relations of production, has nevertheless been made real."

Conclusion

Abstract: *The conclusion specifies the meaning of freedom in the Anthropocene against the background of Chapters 1–3. We then discuss the implications of our book for contemporary environmental politics. The chapter concludes with a discussion of the significance of critical theory in helping us comprehend our current ecological predicament.*

Stoner, Alexander M. and Andony Melathopoulos. *Freedom in the Anthropocene: Twentieth-Century Helplessness in the Face of Climate Change.* New York: Palgrave Macmillan, 2015. DOI: 10.1057/9781137503886.0009.

The purpose of this book has been to clarify the meaning of the Anthropocene by situating it in theoretical, historical, and societal contexts. An important concern of this contextualization has been to elucidate the linkage between the recent popularization of the term and the profound sense of helplessness that comes at the close of the twentieth century regarding the capacity of society to self-consciously transform itself. Our investigation has shown that although the Anthropocene alludes to an underlying historical dynamic driving the opposition between modern society and ecological well-being, it is unable to account for its own very recent turn to history, particularly in relation to large-scale historical dynamics of the twentieth century. Despite the debate generated by the term, the Anthropocene has yet to facilitate a critical and historical understanding of the environment–society problematic wherein societally induced ecological degradation is compounded in proportion to our awareness of these problems. Although the Anthropocene may help us discern humanity's predictable ecological collapse, its account of what generates this transformation ends up linear and without the potential for being otherwise. As a result, the specific type of social transformation the Anthropocene entails, as well as its conditions of possibility, results in the externalization of the factors of transformation.

In this book we have shown how critical theory—especially as represented in the writings of Lukács, Adorno, and Postone—can enable us to think through the problems raised by the Anthropocene, not only as individuals whose lives span the Great Acceleration, but through their commitment to understanding freedom and transformation in relation to its context. The Anthropocene is not something to be dismissed but understood relative to the failure to advance freedom in the twentieth century. The task of critical theory, then, with respect to the mounting ecological problems relative to our current moment, is to critically reflect on this fact; to discern this helplessness as a product of history that might yet be overcome.

We have seen that Lukács' critique of reification, written in 1923, is an attempt to confront the reconfiguration of power and society necessitated by the accumulated failure of revolutionary politics. Lukács directs focus toward the subjective dimension of the commodity form of social relations; that is, how the world appears to those living within capitalist society. Grappling with the tension between the necessity of revolution and the growing role of the state at the turn of the century, Lukács sought to grasp how the subjective dimensions of the worker's movement was constitutive not only of the conscious self-transformation of society, through the revolutionary overcoming of wage labor, but also could reconstitute wage labor in new social forms of domination that would become increasingly naturalized, or "reified." In doing so, Lukács sought to discern the prospects for regaining revolutionary practice from existing reified consciousness within the proletarian movement. Yet, new possibilities for freedom were not actualized, as evidenced by the rise of state planning and the incorporation of the worker's movement. Indeed, the postliberal reconfiguration of business–labor–government relations during the 1930s, mediated by "administered" society, is what allowed for the subsequent ratcheting up of production, which, unleashed through these new organizational forms on a global scale following WWII, gave the Great Acceleration its objective "runaway" character.

Adorno's critique of identity thinking, which he elaborated most fully in 1966, would not have been possible without the rise of authoritarianism and the subsequent and concomitant consolidation of social domination in capital. Whereas the problem of freedom in the context of capitalism took on practical dimensions for Lukács, Adorno, writing at a later stage of capitalist development, is left to critically reflect on the ability to think about the potential for revolution, but during a time when practically overcoming unfreedom is seemingly opaque. Adorno's critique of identity thinking registers the ability to be critical of social structure and consciousness at a time when their own unfulfilled potential persists.

Chapter 2 considered Adorno's critique of identity thinking in relation to the development of the Great Acceleration, and the subjective dimension of the environment–society problematic in particular. We discussed the emergence of contemporary environmental politics in the 1960s and the way in which its opportunity to transform society appears to have been progressively foreclosed throughout the latter half of the twentieth century. Our approach is less linear than the conventional accounts of contemporary environmentalism that explain its rise with reference to technology (see, e.g., Edwards, 1996; 2010) or the spike in environmental degradation following WWII (see, e.g., Gottlieb, 2005 [1994]). Indeed, as discussed in Chapter 2, it is the identity thinking in contemporary environmentalism that forecloses the possibility that its discontents might render ecological degradation recognizable and by extension subject to its free and conscious overcoming by society. Hence, although the environmental effects of the post-WWII period are becoming increasingly visible, solutions to these problems continue to be put forth in terms of the problem itself—and apparently without the potential for being otherwise.[1] Indeed, while many researchers, academics, and activists have seized on the urgency of reducing societally induced environmental degradation, the rate at which such degradation increases continues to accelerate.

The possibility of conceiving something like the Anthropocene is strongly conditioned by the subjectivity expressed by the environmental movement that emerged alongside Adorno's *Negative Dialectics*. The Anthropocene, then, is history from the standpoint of the identity thinking of contemporary environmentalism—a history characterized by unfreedom; hence, its drivers as well as the possibility of freedom and agency remain exogenous to society.[2] The social, political, and cultural mediations (between the non-identical, emancipatory moment in contemporary ecological subjectivity and its actual concrete expression in environmentalism) do not lead to any sense of deep structure but to an even more fragmented sense of the world. The Anthropocene is an attempt at World History that fails because it cannot grasp its movement.

Our analysis of Postone's critique of traditional Marxism in Chapter 3 specified the connection between economic growth and environmental degradation and how environmental degradation is necessarily connected to social domination under capitalism. Significantly, it is during the post-WWII period that the "growth imperative rapidly became a core societal value that drove both the socio-economic and the political spheres" (Steffen et al., 2011: 850). Yet, the term "economic growth" has, amid its widespread usage in political discourse, acquired a meaning independent of the context to which it refers. Similarly, the question of what exactly the so-called growth imperative is has remained opaque. Not surprisingly, the linkage between economic growth and ecological degradation also remains unclear.

In our discussion of Postone's critical Marxian theory outlined in Chapter 3 we suggested a more adequate conceptualization of the Great Acceleration in terms of the unfolding of the contradiction between wealth and value throughout the latter half of the twentieth century. On the basis of Postone's appropriation of Marx's category of socially necessary labor time, we emphasized the "treadmill" of production of value and the growing necessity of accelerating the rate at which biophysical throughput increases. As we have seen, this allows one to specify, in much more rigorous terms, the "runaway" character of the Great Acceleration. Postone's theory is significant in part because it allows for a critical, dynamic, and reflexive account of the form economic growth must take via the production of value in terms of the inextricable connection between the intensifying domination of people by time, on the one hand, and the necessity of increasing biophysical throughput exponentially, on the other.

Chapter 3 also directed focus toward the ways in which Postone's theory resonates with the contemporary world insofar as revolutionary practice has ebbed to such an extent that all that remains is theory. Although Postone uses theory to discern helplessness, he does so by reinterpreting Marx against so-called traditional Marxism after the clear failure of Marxism. In other words, Postone's insight comes with the foreclosing of the practical dimensions that Lukács was able to reflect upon, and which was palpable in his moment. In this sense, Postone's critique of Lukács is significant not in terms of the theoretical differences between the two (as Postone would have it) but because it registers a change in historical circumstances (the 1960s versus 1917).

1 Contemporary environmental politics

As we have discussed, the Anthropocene is characterized by a growing gap between society's awareness of ecological degradation and our ability to change the underlying social structure that gives rise to this degradation (Stoner, 2014). By working through the thought of Lukács, Adorno, and Postone we gain insight into the reasons why contemporary ecological subjectivity has proven so remarkably unable to transform society in the direction of being less ecologically destructive. Adorno, in particular, suggests that the problem resides in the regression of a specific relationship between society and consciousness: the relation of theory to praxis. While for Lukács the deeper structure of society could be grasped theoretically in and through the reified forms of politics that existed among workers, Adorno is able to recognize the disintegration of this relation by the 1960s as new forms of discontent emerged. Yet, even for Lukács reification evinced a decomposition in how Marx's theory (i.e., the insights of the critique of political economy) was being related to proletarian consciousness in the Second International. As Lukács' contemporary Karl Korsch would observe, Second International Marxism broke "the umbilical cord" of the "natural combination" among various components of Marx's theory (Korsch, 1970 [1923]: 59). It was only by virtue of the attempt to (once again) relate theory to praxis by a new generation of Marxists, including Rosa Luxemburg and Lenin, that reification could be productively engaged in the service of provoking a political crisis that would enable the deeper structure of society to be rendered comprehensible, and in turn, changeable.

As Postone's (1978) critique of "traditional Marxism" suggests, the *potential* for environmental politics in the 1970s to recognize itself as a reflection of this deeper social structure existed *in theory*. Clearly, the subsequent antagonism in the 1980s between workers, whose employment was threatened by environmental protection, and environmentalists expresses not only the *need* for such a theory, but the failure to generate forms of environmental politics that could relate such a theory to the prevailing forms of discontents. What is missing, in other words, is not political activity or critical theory per se, but recognition of the necessity to mediate the two. Adorno's insights into identity thinking locate this problem (of the growing inability to relate a theory of totality to forms of existing discontent) as the key missing dimension in the Great Acceleration. Attempts to relate theory to praxis in the 1960s exhibited the "tendency of slipping into the predominance of praxis," since the political movements of the period, including environmentalism, "defame theory as a form of oppression" (Adorno, 1998 [1969]: 290). For Adorno, the subsequent elevation of praxis in the 1960s, as a way to falsely resolve the problem of relating theory to praxis, is conditioned by prior history—namely, the inability to work through the historical failure of the revolutionary politics of Lukács' moment. As Adorno (1998 [1969]: 292) suggests, "it is no coincidence that the ideals of immediate action (...) have been resurrected after the willing integration of formerly progressive organizations that now in all countries of the earth are developing characteristics traits of what they once opposed." The deep confusion over the failure of revolutionary politics at the beginning of the Great Acceleration expresses itself in terms of an "impatience with theory" that further disintegrates the awareness of the need to relate theory and praxis anew (Adorno, 1998 [1969]: 292). Such "impatience" has grown ever more acute given the scale of the threat posed by ecological degradation. Yet, working through the failures of revolutionary politics at the beginning of the Great Acceleration holds the key to understanding why the urgent immediate tasks faced by contemporary environmental politics seem profoundly unable to affect social transformation in the present.

Our approach of relating the critiques of Lukács, Adorno, and Postone across the Great Acceleration enables us to attain an important vantage for considering various contemporary theories on the increasing exhaustion of environmental politics after the 1970s, particularly by posthumanists (see, e.g., Washick et al., 2015; Bennett, 2010) and those who characterize the present as being constituted by post-ecologist politics (e.g., Blühdorn, 2007; 2013). Although these theories are useful in describing the impasse that characterizes contemporary environmental politics, they remain entirely descriptive in their criticism. These insights, in other words, are not critical in that they are unable to locate, in theory, the potential for discontents (described by post-ecology or posthumanism) to point beyond themselves. Although these theories occasionally draw on Adorno (e.g., Bennett, 2010: 1–19), they fail to recognize his key insight, namely that the separation of theory and praxis is strongly conditioned by history, such that any attempt to regain transformative praxis would entail working through the past failures to relate the two. What is lacking at present is the ability to understand the Great Acceleration relative to the changing relation of theory and praxis.[3] Rather than explaining our growing inability to understand—let alone change—the deeper structure of society, contemporary political theory naturalizes the regression in how theory and praxis relate, explaining its current state as an ontological product of matter (posthumanism) or the irreversible results of the failure to advance politics from 1960 to 1980 (post-ecology). Against these approaches we explained the Great Acceleration as the accumulation of failures in attempting to relate theory and practice, which points toward the urgency for grasping the meaning of these historical political failures theoretically. The importance of this dimension was certainly not lost on a revolutionary figure such as Rosa Luxemburg at the beginning of the Great Acceleration. As she would assert: "we are not lost, and we will be victorious if we have not unlearned how to learn" (Luxemburg, 1996 [1915]). The prime political task at this stage in the Great Acceleration is being able to learn again, by learning the lessons of relating theory and practice in history.

2 The necessity of critical theory

Our investigation has shown that understanding the meaning of the Anthropocene as an expression of widespread helplessness requires engaging the environment–society problematic from a historical and critical theoretical perspective. Only in this way can the domination of biophysical nature as well as our inner human nature be elucidated as codetermined by the same sociohistorical process. This, in turn, requires scrutinizing such domination in a radical, immanent manner. For a critical theory of the Anthropocene is made possible by the very same social conditions it critically reflects on and seeks to move beyond. Indeed, confronting the Anthropocene on its own terms, as we have attempted here, means recognizing the social mediations between actual human–ecological transformation, on the one hand, and our social conception and understanding of the natural environment, on the other, without rendering these two dimensions identical.

Situated within the critical theoretical scaffolding provided by Lukács, Adorno, and Postone outlined above, both human–ecological transformation and contemporary ecological subjectivity are rooted in alienation and reification, which, constituted by the capitalist mode of production, must be understood as key *mediating* processes between the subjective and objective dimensions of the environment–society problematic. Drawing on Marx—an insight furthered by Lukács, Postone, and Adorno, albeit in different ways—this theoretical understanding of mediation is also a theory of praxis whereby (*á la* alienation) people create structures through social practice that in turn dominate them[4] while processes of reification (*á la* alienation as "second nature") simultaneously *rewrite reality* so as to inhibit these very same humans from "consciously" recognizing that this is indeed the case.[5] Understood dialectically, this implies that people are both producer and product of alienated socioecological relations of domination.

Combining Lukács' focus on the subjective dimensions of the commodity form with Postone's categorial appropriation of Marx's categories allows one to capture both objective structural drivers of human–ecological transformation as synchronous with the historical development of the commodity form of social relations. Gleaning insight from Adorno, we underscored the reciprocal need of both the subjective and objective dimensions of the environment–society problematic while emphasizing the disproportionate priority of the objective dimension, which, under existing social conditions, operates through individuals as isolated particulars via the unfolding of "the inner composition of elements of nature and elements of history within history itself" (Adorno, 2006 [1964/65]: 116). Our approach therefore illuminates the constellation of social forces which make up a form of abstract (sociobiophysical) domination that effectively operates through social and ecological domination while recognizing these processes as nonidentical to, yet simultaneously shaping the possibilities of subjective experience (Stoner, 2014: 633).

Drawing on Postone's critique of the double character of commodity-determined labor allows for a dynamic reconceptualization of reification in terms of the sociobiophysical tensions underlying the production of value, that is, the necessary social–ecological domination underlying value as the continual necessity of the present, where the continual necessity of the present is recognized as immanently dynamic. That is to say, capital, as self-valorizing value, tends toward accelerating rates of increases in productivity, which in turn necessitates accelerating the rate at which biophysical "throughput" increases, thereby effecting quantitative increases in "material wealth" output, even though the resultant increases in value are effected only indirectly. The increasingly anachronistic character of value as a measure of material wealth *and* as a structural precondition of the capitalist mode of production is precisely the point at which value (as the necessity of the present) points beyond itself.

In this sense, accelerating the rate at which biophysical throughput increases—the form so-called economic growth must take—will remain necessary as long as human and nonhuman natures exist primarily as material bearers of objectified time, that is, to serve the continual expansion of capital. In this sense, the ontology of the material realm—a realm so defended by environmental "realists"—is, quite literally, the ontology of the social *mediation* of time, labor, and social domination, or, in Adorno's words, "the ontology of the wrong state of things" (Adorno, 1973 [1966]: 11).

The question of "wilderness" (see, e.g., Cronon, 1991) and the so-called end of nature (see, e.g., McKibben, 1989) must, then, be formulated in terms of social mediation, which, as we have argued, must be rooted in a critical and historical understanding of the reciprocal dynamism of the subject–object dimensions of the environment–society problematic whereby the (heteronomous) objective dimension takes priority insofar as this heteronomy is *always already* defined (and being defined) as such. Without explicit recognition of this priority, research efforts are at the inherent risk of becoming ideology. For insofar as the priority of the objective dimension is always already defined as heteronomous, failure to recognize this priority as such, and/or failure to make such recognition an explicit and integral part of the activity of social research, *de facto* risks becoming a reflection of the present in reified, alienated form.

This approach, with its understanding of mediation, subjectivity, and abstract socioecological domination, allows one to distinguish the meaning of the Anthropocene. The Anthropocene registers the extent to which the twentieth century was characterized not by freedom, but by the wholesale return of structural constraints that restrict transforming current socioecological relations in more sustainable ways. Yet, the Anthropocene does not critically reflect on its own immersion in this history and therefore unintentionally projects helplessness in the present backward over the past 250 years.

The history of the Anthropocene is "something that is done to people" (Adorno, 2006 [1964/65]: 9). Although the Anthropocene is meant to denote a new epoch in which the activity of human societies has become the determinant force in transforming Earth's planetary systems, humanity is not self-consciously choosing to control such transformation. In modern capitalist society humans are not free to control their control of nature, nor can they control their own lack of control. The problem of freedom, as we have seen, has become obscured with the unfolding of the twentieth century. Today, the term is no longer measured against its context; and yet, there can be no freedom without a subjective, conscious interest (Adorno, 2006 [1965/65]: 6). Hence, the question of the meaning of freedom in the Anthropocene remains a question of "whether we can construct history without committing the cardinal sin of insinuating meaning where none exists" (Adorno, 2006 [1964/65]: 9).

Notes

1   As Dahms (2006: xiii) notes, "The very success of the post-World War II configuration appears to have begun to impair the ability of decision-makers in key institutions and organizations in industrialized societies, to confront, and even to manage, newly emerging challenges and threats today—in ways that would contain, or reduce, their urgency and disruptive nature." Global climate change is a case in point.

2 In the language of Crutzen and his colleagues (Steffen et al., 2007; 2011), the history of the Great Acceleration bears the fragmentation inherent in environmentalism's attempt to mediate discontents with environmental quality—e.g., through neoclassical economists and fossil fuel cars—as discrete aspects of society but never its whole.
3 Throughout this book, we have attempt to demonstrate how the more specific lack of agency associated with contemporary environmental politics is connected to the more general theory/praxis problem relative to the decline of Left politics in the twentieth century. To this end, we think the approach taken by Blumberg and Nagales (2008) provides a good framework for thinking through the historical origins of the current inability to relate theory and praxis in environmental politics.
4 On alienation as self-generated domination, see Postone (1993).
5 On alienation as "second nature," see Dahms (2011).

References

Adorno TW (1978 [1938]) On the Fetish-Character in Music and the Regression of Listening. In Arato A and Gebhardt E (eds) *The Essential Frankfurt School Reader*. New York [Etc.]: Bloomsbury Press, pp. 270–299.

Adorno TW (2003 [1942]) Reflections on Class Theory. In Tiedemannm R (ed.) *Can One Live After Auschwitz?: A Philosophical Reader*. Stanford, CA: Stanford University Press, pp. 93–110.

Adorno TW (1998 [1962]) Progress. In Pickford HW (ed.) *Critical Models: Interventions and Catchwords*. New York: Columbia University Press, pp. 143–160.

Adorno TW (1993 [1963]) *Hegel: Three Studies*. Cambridge, MA and London: The MIT Press.

Adorno TW (2006 [1964/65]) *History and Freedom: Lectures 1964–1965*. Malden, MA and Cambridge: Polity Press.

Adorno TW (2008 [1965]) *Lectures on Negative Dialectics*. Malden, MA and Cambridge: Polity Press.

Adorno TW (1973 [1966]) *Negative Dialectics*. New York and London: Continuum.

Adorno TW (2000 [1968]) *Introduction to Sociology*. Malden, MA and Cambridge: Polity Press.

Adorno TW (2003 [1969]) Late Capitalism or Industrial Society?: The Fundamental Question of the Present Structure of Society. In Tiedemann R (ed.) *Can One Live after Auschwitz?: A Philosophical Reader*. Stanford, CA: Stanford University Press, pp. 111–125.

Adorno TW (1976 [1969]) Introduction. In Adorno TW, et al. (eds) *The Positivist Dispute in German Sociology*. New York: Harper & Row, pp. 1–67.

Adorno TW (1998 [1969]) Resignation. In Pickford HW (ed.) *Critical Models: Interventions and Catchwords*. New York: Columbia University Press, pp. 289–293.
Adorno TW (1997 [1970]) *Aesthetic Theory*. Minneapolis: University of Minnesota Press.
Amadae SM (2003) *Rationalizing Capitalist Democracy: The Cold War Origins of Rational Choice Liberalism*. Chicago and London: The University of Chicago Press.
Anastasovski TE (2014) Geoengineering the Climate: Into the Great Wide Open. *The Economist*. Available at: http://www.economist.com/news/science-and-technology/21635983-scientific-studies-techniques-deliberately-modifying-climate-are (accessed January 23, 2015).
Balmford A and Cowling RM (2006) Fusion or Failure? The Future of Conservation Biology. *Conservation Biology* 20(3): 692–695.
Bennett J (2010) *Vibrant Matter: A Political Ecology of Things*. Durham, NC: Duke University Press.
Berman P (2005) *Power and the Idealists: Or, the Passion of Joschka Fischer and Its Aftermath*. Brooklyn, NY: Soft Skull Press.
Bernstein JM (2004) Negative Dialectic as Fate: Adorno and Hegel. In Huhn T (ed.) *The Cambridge Companion to Adorno*. Cambridge and New York: Cambridge University Press, pp. 19–50.
Biehl J (2012) Bookchin's Trotskyist Decade: 1939–1948. *Platypus Review* 52(1): 3–4.
Biro A (2005) *Denaturalizing Ecological Politics: Alienation from Nature from Rousseau to the Frankfurt School and beyond*. Toronto: University of Toronto Press.
Blühdorn I (2007) Sustaining the Unsustainable: Symbolic Politics and the Politics of Simulation. *Environmental Politics* 16(2): 251–275.
Blühdorn I (2013) The Governance of Unsustainability: Ecology and Democracy after the Post-democratic Turn. *Environmental Politics* 22(1): 16–36.
Blumberg B and Nagales PC (2008) Marx after Marxism: An Interview with Moishe Postone. *Platypus Review* 3. Available at: http://platypus1917.org/2008/03/01/marx-after-marxism-an-interview-with-moishe-postone/ (accessed 8 January 2015).
Blumberg B, Leonard SA, Khan A, Rubin R, and Cutrone C (2009) The Decline of the Left in the 20th Century. *Platypus Review* 17. Available at: http://platypus1917.org/category/pr/issue17-pr/ (accessed 27 January 2015).

Boltanski L and Chiapello (2005) *The New Spirit of Capitalism*. Brooklyn, NY: Verso.

Bookchin M (1986 [1969]) Listen, Marxist! In *Post-Scarcity Anarchism*. Montreal, Canada: Black Rose Books, pp. 193–242.

Brimblecombe P (2006) The Clean Air Act after 50 Years. *Weather* 61(11): 311–314.

Brown AG, et al. (2013) The Anthropocene: Is There a Geomorphological Case? *Earth Surface Processes and Landforms* 38(4): 431–434.

Cain Z and Lovejoy S (2004) History and Outlook for Farm Bill Conservation Programs. *Choices* 19: 37–42.

Carson R (1962) *Silent Spring*. New York: Houghton Mifflin.

Cook D (2006) Adorno's Critical Materialism. *Philosophy & Social Criticism* 32(6): 719–737.

Cook D (2011) *Adorno on Nature*. Minneapolis, MN: Acumen.

Constant B (1988 [1819]) The Liberty of the Ancients Compared with the that of the Moderns. In Fontana B (ed.) *Political Writings*. Cambridge: Cambridge University Press, pp. 309–328.

Coyle L (2011) The Spiritless Rose in the Cross of the Present: Retracing Hegel in Adorno's Negative Dialectics and Related Lectures. *Telos* (155): 39–60.

Cronon W (1991) *Nature's Metropolis: Chicago and the Great West*. New York: W.W. Norton & Company.

Crutzen PJ and Stoermer EF (2000) The Anthropocene. *Global Change Newsletter* 41: 17–18.

Crutzen PJ (2002) Geology of Mankind. *Nature* 415(6867): 23–23.

Crutzen PJ (2006) Albedo Enhancement by Stratospheric Sulfur Injections: A Contribution to Resolve a Policy Dilemma? *Climatic Change* 77: 211–220.

Cutrone C (2013) Adorno's Marxism. PhD Dissertation, University of Chicago.

Cutrone C (2014) When was the Crisis of Capitalism? Moishe Postone and the Legacy of the 1960s New Left. *Platypus Review* 40. Available at: http://platypus1917.org/2014/10/18/crisis-capitalism-moishe-postone-legacy-1960s-new-left/ (accessed December 18, 2014).

Dahms HF (2000) Introduction. In HF Dahms (ed.) *Transformations of Capitalism: Economy, Society, and the State in Modern Times*. New York: Palgrave Macmillan, pp. 1–27.

Dahms HF (2006) Introduction: Globalization between the Cold War and Neo-imperialism. In Lehman J and Dahms HF (eds)

Globalization Between the Cold War and Neo-imperialism: Current Perspectives in Social Theory, Volume 24. Oxford: JAI Press, pp. xi–xvi.

Dahms HF (2011) The Early Frankfurt School Critique of Capitalism: Critical Theory between Pollock's "State Capitalism" and the Critique of Instrumental Reason. In *The Vitality of Critical Theory, Current Perspectives in Social Theory, Volume 28*. Bingley: Emerald Group Publishing Limited, pp. 1–44.

Doel RE (2003) Constituting the Postwar Earth Sciences The Military's Influence on the Environmental Sciences in the USA after 1945. *Social Studies of Science* 33(5): 635–666.

Donnelly M (2014) *Sixties Britain: Culture, Society and Politics*. London: Routledge.

Dykema JA, Keith DW, Anderson, JG, and Weisenstein, D (2014) Stratospheric Controlled Perturbation Experiment: A Small-scale Experiment to Improve Understanding of the Risks of Solar Geoengineering. *Philosophical Transactions of the Royal Society A* 372: 20140059.

Edwards P (1996) *The Closed World: Computers and the Politics of Discourse in Cold War America*. Cambridge, MA and London: The MIT Press.

Edwards P (2010) *A Vast Machine: Computer Models, Climate Data, and the Politics of Global Warming*. Cambridge, MA and London: The MIT Press.

Etheridge DM, et al. (1996). Natural and Anthropogenic Changes in Atmospheric CO_2 over the last 1000 Years from air in Antarctic Ice and Firn. *Journal of Geophysical Research: Atmospheres* 101: 4115–4128.

Feenberg A (1996) The Commoner-Ehrlich Debate: Environmentalism and the Politics of Survival. In Macauley D (ed.) *Minding Nature: The Philosophers of Ecology*. New York: The Guilford Press, pp. 257–282.

Feenberg A (1999) A Fresh Look at Lukács: On Steven Vogel's Against Nature. *Rethinking Marxism* 11(4): 83–93.

Fenger J (1999) Urban Air Quality. *Atmospheric Environment* 33: 4877–4900.

Field CB, Campbell, JE, and Lobell, DB (2008). Biomass Energy: The Scale of the Potential Resource. *Trends in Ecology & Evolution* 23: 65–72.

Fischer J, et al. (2007) Mind the Sustainability Gap. *Trends in Ecology & Evolution* 22(12): 621–624.

Friedmann H (1982) The Political Economy of Food: The Rise and Fall of the Postwar International Food Order. *American Journal of Sociology* 88: S248–S286.

Friedmann H (1993) The Political Economy of Food: A Global Crisis. *New Left Review* I/197: 29–57.

Gaveau DLA, et al. (2014) Major Atmospheric Emissions from Peat Fires in Southeast Asia during Non-drought Years: Evidence from the 2013 Sumatran Fires. *Nature: Scientific Reports* 4: 6112.

Gerasimchuk I and Koh PY (2013) *The EU Biofuel Policy and Palm Oil: Cutting Subsidies or Cutting Rainforest?* Winnipeg, MB: The International Institute for Sustainable Development (IISD).

Ghazoul J, Koh LP, and Butler RA (2010) A REDD Light for Wildlife-friendly Farming. *Conservation Biology* 24(3): 644–645.

Gorz A (1967) *Strategy for Labor: A Radical Proposition.* Boston, MA: Beacon Press.

Gottlieb R (2005 [1994]) *Forcing the Spring: The Transformation of the American Environmental Movement.* Washington, DC: Island Press.

Headey D and Fan S (2008) Anatomy of a Crisis: The Causes and Consequences of Surging Food Prices. *Agricultural Economics* 39: 375–391.

Hegel GWF (1977) *The Phenomenology of Spirit.* Trans. AV Miller. New York [Etc.]: Oxford University Press.

Heyck TW (2008) *The Peoples of the British Isles: A New History. From 1870 to the Present*, 3rd ed. Chicago, IL: Lyceum Books.

Hobsbawm E (1962) *The Age of Revolution: 1789–1848.* London: Weidenfeld and Nicolson.

Hobsbawm E (1994) *The Age of Extremes: The Short Twentieth Century, 1914–1991.*

Hodgson G (1976) *America in Our Time: From World War II to Nixon—What Happened and Why.* New York: Princeton University Press.

Hooks G and McLauchlan G (1992) The Institutional Foundation of Warmaking: Three Eras of U.S. Warmaking, 1939–1989. *Theory and Society* 21: 757–788.

Hooks G and Smith CL (2005) Treadmills of Production and Destruction Threats to the Environment Posed by Militarism. *Organization & Environment* 18(1): 19–37.

Hooks G and Smith CL (2012) The Treadmill of Destruction Goes Global. In Gouliamos K and Kassineris C (eds) *The Marketing of War in the Age of Neo-Militarism.* New York: Routledge, pp. 60–86.

Horkheimer M (1978 [1940]) The Authoritarian State. In Arato A and Gebhardt E (eds) *The Essential Frankfurt School Reader*. New York [Etc.]: Bloomsbury, pp. 95–117.

Huhn T (2004) Introduction: Thoughts Beside Themselves. In Huhn T (ed.) *The Cambridge Companion to Adorno*. Cambridge and New York: Cambridge University Press, pp. 1–18.

Hunt A, et al. (2003) Toxicologic and Epidemiologic Clues from the Characterization of the 1952 London Smog Fine Particulate Matter in Archival Autopsy Lung Tissues. *Environmental Health Perspectives* 111: 1209–1214.

Jarvis S (2004) Adorno, Marx, Materialism. In Huhn T (ed.) *The Cambridge Companion to Adorno*. New York: Cambridge University Press, 79–100.

Joll J (1968) *The Second International, 1889–1914*. London: Weidenfeld & Nicolson Press.

Karlsson R (2013) Ambivalence, Irony, and Democracy in the Anthropocene. *Futures* 46: 1–9.

Keeble D (1978) Industrial Decline in the Inner City and Conurbation. *Transactions of the Institute of British Geographers* 3: 101–114.

Keith DW, Parson E, Morgan, MG (2010) Research on Global Sun Block Needed Now. *Nature* 463: 426–427.

Keith DW, Duren R, MacMartin DG (2014) Field Experiments on Solar Geoengineering: Report of a Workshop Exploring a Representative Research Portfolio. *Philosophical Transactions of the Royal Society A: Mathematical, Physical and Engineering Sciences* 372: 20140175.

Kivisto P (1986) What's New about the "New Social Movements"?: Continuities and Discontinuities with the Socialist Project. *Mid-American Review of Sociology*: 29–43.

Koh LP and Ghazoul J (2008) Biofuels, Biodiversity, and People: Understanding the Conflicts and Finding Opportunities. *Biological Conservation* 141: 2450–2460.

Koh LP and Ghazoul J (2010) Spatially Explicit Scenario Analysis for Reconciling Agricultural Expansion, Forest Protection, and Carbon Conservation in Indonesia. *Proceedings of the National Academy of Sciences* 107: 11140–11144.

Koh LP and Wilcove DS (2007) Cashing in Palm Oil for Conservation. *Nature* 448: 993–994.

Korsch K (1970 [1923]) *Marxism and Philosophy*. New Left Books, London. Translated by Fred Halliday.

Lasch C (1968) Where do we go from here? *New York Review of Books*, October 10, Issue. Available at: http://www.nybooks.com/articles/archives/1968/oct/10/where-do-we-go-from-here/ (accessed January 20, 2015).

Laskin D (2006) The Great London Smog. *Weatherwise* 59: 42–45.

Latour B (2014) Agency at the Time of the Anthropocene. *New Literary History* 45: 1–18.

Lenin VI (1975 [1902]) What Is to Be Done? In Tucker RC (ed.) *The Lenin Anthology*. New York: W.W. Norton & Company, pp. 12–114.

Lipset SM (1997) *American Exceptionalism: A Double-Edged Sword*. London and New York: W.W. Norton & Company.

Lukács G (1971 [1923]) *History and Class Consciousness*. Cambridge, MA: The MIT Press.

Luxemburg R (1970 [1899]) Reform or Revolution. In Waters MA (ed.) *Rosa Luxemburg Speaks*. New York: Pathfinder Press, pp. 33–90.

Luxemburg R (1996 [1915]) The Junius Pamphlet. Luxemburg Internet Archive (marxists.org) Hollis D (trans.). Available at: https://www.marxists.org/archive/luxemburg/1915/junius/index.htm (accessed 10 February 2015).

MacFarling MC, et al. (2006) Law Dome CO_2, CH_4 and N_2O Ice Core Records Extended to 2000 Years BP. *Geophysical Research Letters* 33(14).

Marx K (1988 [1844]) *Economic and Philosophic Manuscripts of 1844*. New York: Prometheus Books.

Marx K (2008 [1852]) *The Eighteenth Brumaire of Louis Bonaparte*. Rockville, MD: Wildside Press.

Marx K (1978 [1852]) The Eighteenth Brumaire of Louis Bonaparte. In Tucker RC (ed.) *The Marx-Engels Reader*, 2nd edition. New York and London: W.W. Norton & Company, pp. 594–617.

Marx K (1974 [1857/58]) *Grundrisse*. New York: Penguin Books.

Marx K (1976 [1867]) *Capital, Volume 1*. New York: Penguin Books.

Marx K (1993 [1871]) The Civil War in France. In Marx K and Lenin VI (eds) *The Civil War in France: The Paris Commune*. New York: International Publishers, pp. 36–85.

Marx K (1981 [1894]) *Capital, Volume 3*. New York: Penguin Books.

Marx K and Engels F (1978 [1848]) Manifesto of the Communist Party. In Tucker RC (ed.) *The Marx-Engels Reader*, 2nd edition. New York and London: W.W. Norton & Company, pp. 469–500.

McGranahan DA, et al. (2013) A Historical Primer on the US Farm Bill: Supply Management and Conservation Policy. *Journal of Soil and Water Conservation* 68: 67A–73A.

McGrew A (1990) The Political Dynamics of the New Environmentalism. *Organization & Environment* 4: 291–305.

McKibben B (1989) *The End of Nature.* New York: Random House.

McLauchlan G (1992) The Advent of Nuclear Weapons and the Formation of the Scientific-Military-Industrial Complex in World War II. In Gregg WB, et al. (eds) *The Military-Industrial Complex: Eisenhower's Warning Three Decades Later.* New York [Etc.]: Peter Lang, pp. 101–128.

McNeill JR (2000) *Something New Under the Sun: An Environmental History of the Twentieth-Century World.* New York: W.W. Norton.

McNeill JR and Unger CR (2010) *Environmental Histories of the Cold War.* New York: Cambridge University Press.

McQuaid K (1994) *Uneasy Partners: Big Business in American Politics 1945–1990.* Baltimore, MD: The Johns Hopkins University Press.

Mewes H (1983) The West German Green Party. *New German Critique*: 51–85.

Mills CW (1956) *The Power Elite.* London, Oxford, and New York: Oxford University Press.

Mukherjee I and Sovacool BK (2014) Palm Oil-based Biofuels and Sustainability in Southeast Asia: A Review of Indonesia, Malaysia, and Thailand. *Renewable and Sustainable Energy Reviews* 37: 1–12.

Murthy V (2009) Reconfiguring Historical Time: Moishe Postone's Interpretation of Marx. In Postone M (ed.) *History and Heteronomy: Critical Essays.* The University of Tokyo Center for Philosophy, pp. 9–30.

Nagourney E (2003) Why the Great Smog of London was anything but great. Available at: http://www.nytimes.com/2003/08/12/science/why-the-great-smog-of-london-was-anything-but-great.html (accessed August 12, 2014).

Nettl P (1965) The German Social Democratic Party 1890–1914 as a Political Model. *Past & Present* 30: 65–95.

Nickell S, Nunziata L, and Ochel W (2005) Unemployment in the OECD since the 1960s: What Do We Know? *The Economic Journal* 115: 1–27.

Nicolaus M (1968) The Unknown Marx. *New Left Review* 1/48: 41–61.

O'Connor B (1999) The Concept of Mediation in Hegel and Adorno. *Bulletin-Hegel Society of Great Britain*, pp. 84–96.

O'Connor B (2004) *Adorno's Negative Dialectic: Philosophy and the Possibility of Critical Rationality.* Cambridge, MA and London: The MIT Press.

O'Connor B (2013) *Adorno.* New York: Routledge.

Olver CH, Shuter BJ, and Minns CK (1995) Toward a Definition of Conservation Principles for Fisheries Management. *Canadian Journal of Fisheries and Aquatic Sciences* 52: 1584–1594.

Opie J (1998) *Nature's Nation: An Environmental History of the United States*. Forth Worth, TX: Harcourt Brace College Publishers.

Oppenheimer C (2003) Climatic, Environmental and Human Consequences of the Largest Known Historic Eruption: Tambora Volcano (Indonesia) 1815. *Progress in Physical Geography* 27: 230–259.

Orr J (2006) *Panic Diaries: A Genealogy of Panic Disorder*. Durham, NC and London: Duke University Press.

Page SE, et al. (2002) The Amount of Carbon Released from Peat and Forest Fires in Indonesia during 1997. *Nature* 420: 61–65

Post J (1977) *The Last Great Subsistence Crisis in the Western World*. Baltimore, MD: The Johns Hopkins University Press.

Postone M (1978) Necessity, Labor, and Time: A Reinterpretation of the Marxian Critique of Capitalism. *Social Research* 45(4): 739–799.

Postone M (1993) *Time, Labor, and Social Domination: A Reinterpretation of Marx's Critical Theory*. New York: Cambridge University Press.

Postone M (1999) Contemporary Historical Transformations: Beyond Postindustrial Theory and Neo-Marxism. *Current Perspectives in Social Theory, Volume 19*. Stanford, CT: JAI Press, Inc., pp. 3–53.

Postone M (2003) Lukács and the Dialectical Critique of Capitalism. In Albritton R and Simoulidis J (eds) *New Dialectics and Political Economy*. New York: Palgrave Macmillan, pp. 78–100.

Postone M (2006). History and Helplessness: Mass Mobilization and Contemporary Forms of Anticapitalism. *Public Culture* 18(1): 93–110.

Postone M (2009) *History and Heteronomy*. The University of Tokyo Center for Philosophy. Available at: http://utcp.c.u-tokyo.ac.jp/publications/2009/06/history_and_heteronomy_critica/index_en.php (accessed 28 January 2015).

Potter C (1998) *Against the Grain: Agri-environmental Reform in the United States and the European Union*. Oxfordshire: CAB International.

Potter C (2009) Agricultural Stewardship, Climate Change and the Public Goods Debate. In Winter M and Lobley M (eds) *What is Land For?: The Food, Fuel and Climate Change Debate*. London: Earthscan, pp. 247–261.

Røpke I (2005) Trends in the Development of Ecological Economics from the Late 1980s to the Early 2000s. *Ecological Economics* 55(2): 262–290.

Rose G (1976) How is Critical Theory Possible? Theodor W. Adorno and Concept Formation in Sociology. *Political Studies* XXIV(1): 69–85.

Rose G (1979) *The Melancholy Science: An Introduction to the Thought of Theodor W. Adorno.* New York: Columbia University Press.

Ruddiman WF (2014) The Anthropocene. *Annual Review of Earth and Planetary Science* 41:45–68.

Sanderson JB (1961) The National Smoke Abatement Society and the Clean Air Act (1956). *Political Studies* 9: 236–253.

Scarrow HA (1972) The Impact of British Domestic Air Pollution Legislation. *British Journal of Political Science* 2: 261–282.

Schnaiberg A (1980) *The Environment: From Surplus to Scarcity.* New York: Oxford University Press.

Scott A (1990) *Ideology and the New Social Movements.* London: Unwin Hyman.

Smith VC (2014) Volcanic Markers for Dating the Onset of the Anthropocene. *Geological Society, London, Special Publications* 395: 283–299.

Steffen W (2008) Looking Back to the Future. *Ambio: A Journal of the Human Environment* 37: 507–513.

Steffen W, et al. (2007) The Anthropocene: Are Humans Now Overwhelming the Great Forces of Nature. *Ambio: A Journal of the Human Environment* 36(8): 614–621.

Steffen W, et al. (2011) The Anthropocene: Conceptual and Historical Perspectives. *Philosophical Transactions of the Royal Society A* 369(1938): 842–867.

Stone A (2006) Adorno and the Disenchantment of Nature. *Philosophy and Social Criticism* 32(2): 231–253.

Stoner AM (2014) Sociobiophysicality and the Necessity of Critical Theory: Moving beyond Prevailing Conceptions of Environmental Sociology in the USA. *Critical Sociology* 40(4): 621–642.

Stothers RB (1984) The Great Tambora Eruption in 1815 and Its Aftermath. *Science* 224: 1191–1198.

Straka, TJ (2009) Evolution of Sustainability in American Forest Resource Management Planning in the Context of the American Forest Management Textbook. *Sustainability* 1: 838–854.

Taylor PJ and Buttel FH (1992) How Do We Know We Have Global Environmental Problems? Science and the Globalization of Environmental Discourse. *Geoforum* 23(3): 405–416.

Tiedemann R (2006) Editor's Foreword. In Adorno TW (ed.) *History and Freedom: Lectures 1964–1965*. Malden, MA and Cambridge: Polity Press, pp. xii–xix.

Tyner WE (2008) The US Ethanol and Biofuels Boom: Its Origins, Current Status, and Future Prospects. *BioScience* 58: 646–653.

Vogel S (1996) *Against Nature: The Concept of Nature in Critical Theory*. New York: SUNY Press.

Washick B, et al. (2015) Politics that Matter: Thinking about Power and Justice with the New Materialists. *Contemporary Political Theory* 14: 63–89.

Webber MJ and Rigby DL (1996) *The Golden Age Illusion: Rethinking Postwar Capitalism*. New York: The Guilford Press.

Zalasiewicz J, Kryza R, and Williams M (2014) The Mineral Signature of the Anthropocene in Its Deep-Time Context. *Geological Society, London*, Special Publications 395: 109–117.

Zalasiewicz J and Williams M (2008) Are We Now Living in the Anthropocene? *GSA Today* 18(2): 4–8.

Zerefos C, et al. (2007) Atmospheric Effects of Volcanic Eruptions as Seen by Famous Artists and Depicted in Their Paintings. *Atmospheric Chemistry and Physics* 7: 4027–4042.

Index

abstract labor, 78, 80, 81, 83, 84, 85, 88, 96n8
abstract time, 84–5, 87–8, 95n6, 95n7, 98n24
Adorno, T. W., 24–6, 49–59, 64, 69, 69–70n1, 70n4, 70n6, 74, 76, 94n1, 98n26, 100–8
"administered" society, 58–62, 76
agricultural overproduction, 8–9, 13
air quality, 3–4
air pollution, 4, 5–6, 10–11
alienation, 22, 40, 77, 78, 95n5, 96n12, 106
Anthropocene, 14, 17n19, 27n3, 33, 41–3, 49, 58, 69, 74, 86, 90, 94, 100, 102, 106, 108
  freedom and critique in, 24, 32
  Great Acceleration phase of, see Great Acceleration
  as helplessness, 90–4, 100, 106
  recognition of, 19–20, 26n1, 30

biodiesel, 7, 8, 9, 15n8, 17n12
biofuel, 7–9, 16n8
  blending standards, 9, 17n13
  crops, 8–9, 17n13, 17n14
biophysical throughput, 86, 90, 97n18, 103, 107

biophysical transformations, 12–14
Blühdorn, I., 66, 68
Bonapartism, 29–30, 44n2, 91
bourgeois consciousness, 33, 36–7, 40, 71n7
bourgeois society, 12, 36, 38, 41, 47n13, 49, 52
bourgeois thought, 38–9
bourgeoisie "freedom", 36–7
Business Committee (BC), 60
business–labor–government relations, 59–62, 101

capital, 25, 29, 35–7, 44n2, 53, 62, 69, 74–5, 76, 79, 82–3, 88–91, 96n11, 96n12, 96n13, 97n15, 97n17, 97–8n20, 98n24, 101
  as self-valorizing value, 84–7, 107
Capital, 82
capitalism, 33, 35–7, 41, 43, 54, 66, 74, 76–8, 81–3, 87–8, 91–2, 101, 102
  finance, 59
  fully developed, 84, 86, 89
  global, 22, 62, 69, 76, 108n1
  late, 58
  liberal, 29
  neoliberal, 25
  state, 25, 32, 44n1, 49, 93–4
  state-centric, 69, 75
Churchill, W., 4

Clean Air Act, 4–5, 15n4, 92
climate change, 69
  global, 63, 64, 79, 108n1
  inaction to, 27n3
  solutions to, 13–14
  technology's role in awareness of, 63–4
coal, 3, 5–6, 15n3, 30
cognitive utopia, 54–5
Cold War, 60, 62–4, 71n14, 72n18
Cold War technology, 63–4, 69
Committee on Economic Development (CED), 60
commodity fetishism, 33–5, 37, 46n10
commodity form, 32–5, 42–3, 46n11, 77, 79, 83, 88, 95n7, 96n14, 100, 106
concepts, 50, 52–4, 58, 70n4
concrete labor, 78, 80, 81, 85, 88, 95n6, 95–6n8
concrete time, 87, 88, 95n7
Constant, B., 12
consumerism, 49, 66, 76, 93
contemporary ecological subjectivity, 22–3, 26, 57, 62–5, 69, 72n17, 91–3, 102, 103, 106
contemporary environmental politics, 6–7, 60, 65, 101, 103–5, 109n3
contemporary environmentalism, 26, 58, 61–6, 69, 91–4, 94n3, 101–2, 104
  critique of, 62–9
critical theories, 25–6, 32–41, 49–50, 52–5, 58, 62–9, 74, 76, 100–1
critique of identity thinking, 25–6, 49–50, 52–5, 58, 74, 76, 101
critique of reification, 32–41, 100–1
Crutzen, P. J., 13–14, 24, 26n1, 27n2, 27n3, 42–3, 109n2

Darkness (poem), 10, 12
determinate negation, 51, 70n4
dialectics, 50–2, 70n6
  see also negative dialectics

Earth Day, 65–7
ecological degradation, 14, 23, 57, 62, 68, 74, 79, 100, 101–2, 103, 105
  see also environmental degradation
ecological subjectivity, see contemporary ecological subjectivity
economic growth, 26, 61, 71n14, 74, 86, 90, 97n19, 102–3, 107
emissions, 3
  carbon/ CO_2, 9, 13
  greenhouse gas (GHG), 6–10, 13, 17n14
environment–society problematic, 22–3, 42, 57–8, 62, 72n17, 75, 79, 95n5, 100, 101, 106–7
environmental degradation, 20–3, 27n3, 59, 61–2, 69, 89, 90, 101–2
  see also ecological degradation
environmental discontent, 58, 61–2, 64–6, 69, 93
environmental politics, see contemporary environmental politics
Environmental Protection Agency (EPA), 66
environmentalism, see contemporary environmentalism

farm income, 8–9, 16n9, 16n10
Feenberg, A., 65
fires, 6–7, 10, 15n6, 15n7, 15–16n8
freedom, 12–13, 20, 23, 30, 32, 36–7, 42, 49, 50, 53, 55–6, 75, 86, 93, 94, 108
French Revolution, 12, 36

Germany, 6, 12, 15n4, 32, 40, 44n1, 44n4, 45n5, 49, 68
Great Acceleration, 20–1, 24–6, 27n3, 27n4, 52, 69, 109n2
  and critique of identity thinking, 25–6, 49–50, 52–5, 58, 74, 76, 101
  and critique of reification, 32–41, 100–1
  and critique of traditional Marxism, 74–9, 90, 93, 104
  and helplessness, 90–4, 100, 106
  reification and the dialectical genesis of, 42–3

Great Acceleration – *Continued*
  social democracy and, 29–33
  socioecological domination and production of value, 82–6
  and the "treadmill dynamic" of labor/time, 84–5, 97n16, 102–3
  value and material wealth, 79–82
  value as the continual necessity of the present, 86–90
the Great London Smog, 2–6, 72n16
  activities that increased, 3
  consequences of, 3–4
  and decline of coal consumption, 5
  response of society to, 3–5
  and social transformation, 12–13
green consumerism, 66, 93
Green Party, 66, 68
greenhouse gas (GHG) emissions, 6–10, 13, 17n14
  see also oil palm, as biofuel
Grundrisse rise der Kritik der politischen Ökonomie, 76–7

Hegel G. W. F., 50–2, 54–5, 70n2, 70n3, 70n5, 70n6
helplessness, 90–4, 100, 106
historical time, 87–9, 92, 98n24
historical transformation, 41, 45n7, 50, 69, 86
History and Class Consciousness (HCC), 25, 29, 32, 40, 41, 46n8, 46n10, 71n7
history, negative philosophy of, 55–7
Holocene, 19, 41, 43
human–ecological transformation, 21, 22, 26, 42, 57–8, 83, 106
human labor, 22, 35, 84, 95n6
Hunt, A., 3

identity thinking, 24, 53, 54, 58, 62, 69, 101, 102, 104
  see also critique of identity thinking
Indonesia, 6–7, 9, 10, 15n7
Industrial Revolution, 19, 21, 24, 29–30, 42, 80
Italy, 12, 17n15, 40

Korsch, K., 31, 45n4, 46n8, 47n16, 71n7, 104

labor, 44n2, 60, 61
  abstract, 78, 80, 81, 83, 84, 85, 88, 96n8
  in capitalist society, 75–9
  commodity-producing, 78, 82–3
  concrete, 78, 80, 81, 85, 88, 95n6, 95–6n8
  fragmentation of, 35–6
  human, 22, 35, 84, 95n6
  industrial, 35
  markets, 29
  unions, 43
  wage, 25, 32, 36, 91, 92, 101
  see also proletariat
Lenin, V. I., 32, 40, 46n8, 46n9, 47n13, 104
liberalism, 60, 64, 71n14
London, 2–6
Lukács, G., 24–5, 29, 32–3, 35, 38, 39, 41, 42, 46, 49, 71n7, 74, 100, 104, 106
Luxemburg, R., 31–2, 40, 45–6n8, 47n13, 104, 105

M-C-M (Money-Commodity-Money), 83–4, 97n15
Malaysia, 7, 9
Marx, K., 22, 25, 26, 29–34, 39–41, 43, 44n2, 44–5n4, 45n7, 46n10, 47n13, 52–3, 70–1n7, 71n9, 74–86, 88, 91, 95–6n8, 96n9, 96n12, 97n15, 103–4, 106
Marxism, 24, 26, 29, 46n9, 49, 60, 70n6
  failure of, 75, 103
  see also traditional Marxism
material wealth, 77, 79–82, 84, 86, 89, 90, 95n6, 96n14, 97n20, 107
mediation, 22, 32, 34, 37, 39, 42, 50, 56–7, 64, 68, 87, 102, 108
  see also social mediation
mortality figures, 3, 10
Mount Tambora, 10, 11, 13, 14, 17n15, 17n19

negative dialectics, 49–52, 54–5, 58, 69n1
Negative Dialectics, 25, 49, 52, 69–70n1, 102
Nelson, G., 65, 67
neoliberalism, 76, 92, 93
Nixon, R. (President), 16n9, 65–6, 92–3
non-identity thinking, 54–5

OECD countries, 21, 98n26
oil palm, as biofuel, 7, 9, 15–16n8
oilseed crops, 7, 8, 9

particulate matter, 3, 4, 6, 10–12
pollutants, 3, 5–6
  see also coal; particulate matter; sulfur; sulfur dioxide
pollution, 72n16
  agents causing, 3, 5–6
  effects of, 3–4, 6
  particulate, 6, 10–12
  regulation of, 3–5, 9, 13–14
  see also air pollution
positivism, 56
Postone, M., 24–6, 46n11, 74–94, 100, 102–7
post-WWII, 4, 6, 8, 14, 16n9, 20, 26, 27n4, 49, 52, 58–62, 72n19, 74, 92, 101–2
pragmatic identity-thinking, 54
proletarian politics, 30–1, 37, 40–1, 43, 91
proletariat, 29–31, 33, 36–7, 40–1, 43, 44n2, 44n3, 46n8, 46n9, 47n13, 47n17, 49, 70–1n7, 96n12, 98n26

rational identity-thinking, 54
rationalization, 33–5, 46n11
reification, 33, 37, 40–3, 47n13, 53
  see also critique of reification
revisionism, 32, 45–6n8
Rose, G., 54, 71n10

satellite photography, 63, 64

Second International, 29, 31–2, 37, 40, 43n1, 45n4, 47n16, 49, 62, 104
Singapore, 6–7
Smith, A., 13, 29, 30
smog, *see* the Great London Smog; Southeastern Asian Haze
social crisis, 8, 29, 31, 36, 43, 61–2, 91, 93
Social Democratic Party (SPD), 31, 43–4n1, 44n4, 45n6, 45n8
social domination, 25, 58, 74, 75, 90, 101, 102, 107
social mediation, 33, 35, 49, 59, 75, 78, 81, 83, 91, 95n5, 106, 107
social transformation, 4–5, 12–13, 20, 25, 43, 62, 75, 87, 100, 105
socially necessary labor time, 79–81, 84–5, 87, 90, 98n22, 102
Solar-Radiation Management (SRM) technologies, 11, 13–14
Southeastern Asian Haze, 6–10, 12–13
state capitalism, 25, 32, 44n1, 49, 93–4
Steffen, W., 20, 21, 24, 27n2, 27n3, 27n4, 97n19
stratosphere, 10–11, 13–14
sulfur, 10–14
sulfur dioxide, 3, 5, 14
surplus value, 81, 83–6, 90, 95n4, 96n10, 97n18
systems analysis, 64

traditional Marxism, 24, 26, 83, 94, 94n1, 102, 103
  critique of, 74–9, 90, 93, 104
Tugan-Baranovsky, M., 37, 40, 41

unemployment, 5, 15n5, 16n9, 29, 45n5, 92, 98n26
United Kingdom, 4–5, 10, 15n4
United States, 4, 6, 7, 9, 11, 12, 15n4, 16n9, 16n10, 61, 64, 65, 92
use-value, 34, 53, 71n9, 78, 79, 80, 82–5, 87–9, 95–6n8

utopian identity-thinking, 54

value, 53, 66
    as the continual necessity of the present, 86–90
    and material wealth, 79–82
    as a measure of wealth, 76–9
    socioecological domination and production of, 82–6
    surplus, 81, 83–6, 90, 95n4, 96n10, 97n18

use-, 34, 53, 71n9, 78, 79, 80, 82–5, 87–9, 95–6n8
volcanoes/volcanic eruptions, 10–12, 13, 17n15, 17n19

wage labor, 25, 32, 36, 91, 92, 101
Watt, J., 19, 20
Weber, M., 33, 34, 46n11
working class, *see* proletariat
WWI, 29, 31–2, 40, 74
WWII, *see* post-WWII

CPSIA information can be obtained
at www.ICGtesting.com
Printed in the USA
LVHW111435161218
600663LV00008B/79/P